The Microbiology and Epidemiology of Infection for Health Science Students

The Microbiology and Epidemiology of Infection for Health Science Students

Peter Meers

Formerly Associate Professor
Microbiology Department
National University of Singapore

Judith Sedgwick

Infection Control Nurse
South Buckinghamshire NHS Trust
High Wycombe, UK

and

Margaret Worsley

Senior Nurse Manager
North Manchester Healthcare NHS Trust
Manchester, UK

CHAPMAN & HALL

London · Glasgow · Weinheim · New York · Tokyo · Melbourne · Madras

Published by Chapman & Hall, 2–6 Boundary Row, London SE1 8HN, UK

Chapman & Hall, 2–6 Boundary Row, London SE1 8HN, UK

Blackie Academic & Professional, Wester Cleddens Road, Bishopbriggs, Glasgow G64 2NZ, UK

Chapman & Hall GmbH, Pappelallee 3, 69469 Weinheim, Germany

Chapman & Hall USA, One Penn Plaza, 41st Floor, New York NY 10119, USA

Chapman & Hall Japan, ITP-Japan, Kyowa Building, 3F, 2-2-1 Hirakawacho, Chiyoda-ku, Tokyo 102, Japan

Chapman & Hall Australia, Thomas Nelson Australia, 102 Dodds Street, South Melbourne, Victoria 3205, Australia

Chapman & Hall India, R. Seshadri, 32 Second Main Road, CIT East, Madras 600 035, India

Distributed in the USA and Canada by Singular Publishing Group Inc., 4284 41st Street, San Diego, California 92105

First edition 1995

© 1995 Peter D. Meers, Judith A. Sedgwick and Margaret A. Worsley

Typeset in Times 10/12pts by Mews Photosetting, Beckenham, Kent
Printed in England by Clays Ltd, St Ives plc

ISBN 0 412 59900 7   1 56593 350 8 (USA)

A catalogue record for this book is available from the British Library

Library of Congress Catalog Card Number: 94-69923

∞ Printed on permanent acid-free text paper, manufactured in accordance with ANSI/NISO Z39.48-1992 and ANSI/NISO Z39.48.1984 (Permanence of Paper).

Contents

Preface

Florence Nightingale set herself to improve hospitals and raise the standard of nursing. A combination of administrative skill, dogged single-mindedness and an autocratic refusal to accept defeat swept obstacles aside and her remarkable achievements were widely acclaimed in her own lifetime. She was over 60 when the science of microbiology began to throw light on the causes of infections. She would not accept the new ideas, and poked fun at what she called the 'germ fetish'. She believed disease was caused by dirt and that it was banished by cleanliness and fresh air.

She was wrong about the germs. Worldwide, the infections they cause kill more people than anything else. Even in countries where cancer and heart disease have become the major causes of death it is still often an infection that tips terminal morbidity into mortality. A survey made some time ago showed that, on average, about one hospital patient in five had an infection, no matter what else was wrong with them. Half of these infections were actually acquired in hospitals, places that Florence Nightingale declared 'should do the sick no harm'.

At the end of the nineteenth century many influential people thought that infection had been banished by the advances in hygiene made in the previous 50 years. The same idea reappeared in the middle of the twentieth century when leading professionals concluded that infection had finally been vanquished by antimicrobial drugs and vaccines. These were false dawns, and as the twenty-first century approaches infection is still a major problem. Microbes may have lost some battles, but they have won others. The war is far from over.

In warfare it is important to understand your enemy. For most of those who enter the caring professions infection is studied as a small part of an overcrowded curriculum. If the amount of time they spend on the subject is influenced by the idea that infection is no longer important we hope this book will serve to correct the error. It has been designed to help newcomers in their initial studies and then continue to be a useful source of reference as they develop in their chosen vocations.

In today's global village infection is an international subject. It is no longer possible to ignore diseases found in distant parts or even those that seem to be only of historic interest. The next jumbo jet can bring an unusual infection to any doorstep or some new disaster rekindle the embers of a barely remembered pestilence. For these reasons due weight has been given to subjects that might, at first, appear exotic. We hope this approach will make the book useful to people who work anywhere in the world.

Everyone concerned with health can contribute something to the diagnosis, treatment or control of infection. They will not succeed without some knowledge of germs (microbiology), of what these can do to people and how to deal with the result (the study of infection), or of the way infections spread and how to prevent this (epidemiology). Although diagnosis, treatment and control may be thought of separately, to be effective in one area requires insight into the others. Our aim has been to provide the overall view that is needed.

P.D.M.
J.A.S.
M.A.W.
1994

Glossary and abbreviations

Words and phrases with more general meanings are defined here in the context of infection or of medical microbiology. Additional definitions may be traced through the Index.

Aerobe A microbe that grows in the presence of oxygen and not in its absence.

AFB Acid-fast bacillus.

AIDS Acquired immune deficiency syndrome.

Anaerobe A microbe that cannot grow in the presence of oxygen.

Antibody A protein molecule produced by the body as a result of immunization with an antigen, that unites with its antigen in a highly specific way.

Antigen Any substance which, when introduced into the body, produces an immune response.

Antimicrobial drug A drug used to treat or prevent an infection (antimicrobial therapy, the use of such a drug).

AZT Azidothymidine, zidovudine; used to treat AIDS.

Bacillus A rod-shaped bacterial cell; also the name of a bacterial genus.

Bacteriology The science of the study of bacteria.

CAI Community-acquired infection.

Cell The basic unit of all living things; may be single or gathered together in groups to form more complex organisms.

Cellular immunity Functions of the immune system that depend on the direct action of specialized cells on foreign materials in the body.

CMV Cytomegalovirus.

Coccus A spherical bacterium.

Colonization A long-term relationship in which a microbe lives on (or in), a host, without any reaction by the host to its presence.

Colony A collection of microbes of the same type, in one place, often grown artificially in a laboratory to form a visible clump.

Commensal A microbe that establishes a colonization without harm to its host.

CNS Central nervous system.

CSF Cerebro-spinal fluid.

Culture A process by which microbes are induced to multiply artificially: (culture medium, a nutrient mixture in or on which culture is performed in a laboratory).

CVS Cardiovascular system.

Deoxyribonucleic acid One form of nucleic acid with a unique chemical structure, found in all living cells; the repository of genetic information.

Dinoflagellates A group of mainly single-celled algae or protozoa equipped with flagella found in freshwater or seawater: they may produce powerful toxins.

DNA Deoxyribonucleic acid.

EBV Epstein-Barr virus.

EM Electron microscope.

Embolus A particle of tissue, bloodclot or pus carried in the bloodstream to reach a site in the body at some distance from its origin.

Endotoxin A toxic constituent of a living microbe that is an integral part of its structure.

Entomology The science of the study of insects and other arthropods.

Enzyme A protein that acts as a biological catalyst in the process of metabolism to effect the breakdown or the build-up of the molecular components of cells.

Epidemiology The science of the study and control of epidemic diseases.

Eukaryote Living organisms whose cells contain DNA packaged into discrete nuclei.

Exotoxin A toxin produced actively by a microbe and secreted into its environment.

Flora The assembly of plants or microbes that are peculiar to a particular region or to a more restricted environment.

Gene A short stretch of DNA, part of a much longer molecule. A gene is the blueprint that determines the precise structure of a molecule in a cell, and it is also concerned with the inheritance of this function.

Genetics The science of the study of genes and heredity.

Genus A collection of species with sufficient overall similarity in their detailed structure for them to be classified together.

HAI Hospital-acquired (nosocomial) infection.

HBV Hepatitis B virus.

HCW Health care worker.

Helminth A worm, (**-ology**) the science of the study of worms.

HIV Human immunodeficiency virus, the cause of AIDS.

Host The larger partner in a host-parasite relationship or colonization: in medical usage, a human being.

HSV Herpes simplex virus.

Humoral immunity Functions of the immune system that depend on substances dissolved in body fluids, particularly those concerned with antibodies.

Iatrogenic infection An infection that is the result of medical treatment.

ICC Infection Control Committee.

ICD Infection Control Doctor.

ICN Infection Control Nurse.

ICT Infection Control Team.

Incubation The exposure of a microbial culture to a temperature and gaseous environment that is optimum for growth; (incubator, the apparatus that provides this environment in the laboratory).

Infection The outcome of an interaction between a host and a microbe such that the host reacts in some observable way. The result is a clinical infection when the individual becomes ill, or a subclinical infection if the evidence for it can only be detected in a laboratory.

Immune (Immunity) The state of an individual who has been immunized or is otherwise insusceptible to infection.

Immune response The reaction of the body's immune system when it is stimulated by some foreign material (antigen).

Immune system Specialized tissues and their chemical products that operate throughout the body to produce various immune responses.

Immunize (Immunization) The process by which an immune response is generated to an antigen introduced naturally or artificially.

Immunocompetent The state of an individual whose immune system is fully operative.

Immunodeficient The state of an individual whose immune system is, to a variable extent, defective.

Immunodepression A reduction in the efficiency of the immune system, due to any cause.

Immunoglobulin(s) Specialized protein(s), widely distributed in body fluids. Nearly all of them are antibodies.

Immunology The science of the study of immunity.

Immunosuppression The production of immunodeficiency by artificial means (drugs, radiation, etc.).

Lysis The dissolution of a microbe as a result of an attack on its cell wall or limiting membrane.

Metabolism The sum of the chemical activities in a cell that provide for its construction, repair and function, together with the production of the energy necessary for these activities.

MIC Minimum inhibitory concentration.

Microbe A living organism or a virus that is too small to see with the naked eye.

Microbiology The science of the study of microbes.

MID Minimum infectious dose.

Molecule The smallest collection of atoms into which a chemical substance can be divided without losing its special identity as a compound.

MRSA Methicillin-resistant *Staphylococcus aureus*.

Mycology The science of the study of fungi.

Normal flora The collection of microbes usually present as colonists in any specified environment, including the surface of the human body.

Nosocomial infection Synonymous with HAI.

Nucleic acid A long string of individual nucleotide sugars (see DNA and RNA).

Nucleus A discrete structure that contains DNA within a nuclear membrane, inside a eukaryotic cell.

Nutrient(s) Foodstuff(s), including those required by microbes.

Papule A small area of skin that can be seen or felt to be raised above its surroundings, caused by some pathological process: a common component of a rash.

Parasite A microbe or some other small creature that lives on or in a host, that derives benefit from the association.

Parasitology The science of the study of parasites, protozoa, helminths and some arthropods. Although bacteria, fungi and viruses may also be parasites it is traditional to exclude them from the subject of parasitology.

Pathogen A microbe or larger parasite that causes harm to its host and so produces infection or disease.

PBP Penicillin binding protein.

Phagocyte A cell whose special function is to engulf and destroy any foreign material found in the body, in particular PML and macrophages or monocytes.

PML Polymorphonuclear leucocyte.

Privileged surface A surface of the body equipped with mechanisms that keep it free of microbes (e.g., those of the healthy respiratory or urinary tracts).

Prokaryote A microbe whose nuclear DNA is not enclosed within a special membrane to form a eukaryotic nucleus. All of the bacteria are prokaryotes.

Protozoology The science of the study of protozoa.

Pustule A small blister or vesicle that contains pus; a constituent of some rashes on skin and mucous membranes.

Ribonucleic acid A form of nucleic acid with a particular chemical structure. RNA is involved in protein synthesis though in some viruses it replaces DNA as the repository of genetic information.

RNA Ribonucleic acid.

Species A basic division within the classification of living organisms: members of the same species have a very high degree of structural and genetic similarity.

Spirochaete A rod-shaped bacterium that adopts a spiral or corkscrew shape.

Spore A resting form of certain microbes that is unusually resistant to desiccation and chemical attack: an adaption to survive adverse conditions.

Symptom A perceptible change in the function or structure of the body that results from injury or disease.

Syndrome A combination of symptoms and signs that are characteristic of a particular form of disease.

Toxin A powerful poison produced by a microbe or some other living thing.

UTI Urinary tract infection.

uV Ultraviolet (light).

Vaccination The process of immunization with a vaccine.

Vaccine A preparation of antigen(s) used to produce immunity. The antigens in vaccines are usually derived from, or consist of, microbes.

Vesicle A raised area with a small blister that contains a clear fluid: a component of some rashes on skin and mucous membranes.

Vibrio A curved Gram-negative rod-shaped bacterium.

Virology The science of the study of viruses.

VZV Varicella-zoster virus.

YF Yellow fever.

ZIG Zoster immune globulin.

Introduction

Overview

1

This chapter is intended for newcomers to the study of the microbiology and epidemiology of infection. In later chapters the subject is divided into a number of separate topics, but certain ideas, words and phrases run through all of them. This overview introduces these basic concepts and it also provides an outline into which the reader can fit the pieces of what would otherwise be a difficult jigsaw puzzle. A brief history is included to fix the subject in time, and show how it developed.

1.1 SOME BASIC FACTS

The first living things to appear on earth would have been too small to see with the naked eye. Today they would be classified as microbes. We may never know how these early microbes worked, but we can deduce that they collected and then assembled some of the simple molecules that surrounded them to form the more complex ones of which they were made. The process required energy, so this must have been provided as well. The word **metabolism** describes these mutually dependent chemical activities.

Living organisms are separated from inanimate objects by the possession of a metabolism and an ability to reproduce. Successive generations appear as copies of their parents. This is made possible by the genetic material carried by each cell. Genetic material is made up of individual **genes**, each of which carries the instructions for the manufacture of one component of the cell. Genes are strung out along filamentous, intricately coiled molecules of nucleic acid. **Nucleic acids** are of two kinds, **ribonucleic acid (RNA)** or **deoxyribonucleic acid (DNA)**. All living cells use DNA as their genetic memory, and RNA as the machinery that converts that memory into working molecules. The DNA of animal and plant cells is wrapped up inside a membrane to form a separate structure, the **nucleus**. Bacteria do not package their DNA. Although they possess nuclear material, it is not identifiable as a nucleus. Viruses are even more different. Not only do they lack the ability to metabolize but each possesses only one kind of nucleic acid. Some use DNA as their genetic memory and others use RNA.

Microbes may or may not have originated in a single common ancestor, but after the passage of millions of years many different kinds have evolved. They have developed so as to be able to populate most of the environments that exist on earth, including the surface of the human body. The science of microbiology is concerned with the study of these microscopic forms of life.

As microbes spread and became more common they began to compete with each other. This may have started when one of them developed a better way of collecting nutritionally important molecules. If these were scarce a microbe that was able to seize a vital morsel ahead of its neighbours had an advantage. When this ability was inherited the progeny of the more successful microbe came to dominate any environment in which the particular nutrient was in short supply. This continued until some other microbe learned an even better way of doing the same thing.

It was not a large step from the production of the special chemicals necessary to achieve an increased nutritional efficiency to the elaboration of others that were more directly harmful to potential rivals. These poisonous compounds were used to disable or kill other microbes. Chemical warfare was not invented by the human race! The more adaptable of the victims soon learned how to defend themselves. The different kinds of microbes that populate most accessible parts of the world today, including the human body, coexist in a state of armed neutrality with frequent territorial scuffles. Penicillin is only one of the many chemical agents that were used by microbes long before they were 'discovered' by humans and made into **antimicrobial drugs**.

A new era started when some microbes found that it was to their advantage to collect together in adherent clusters of **colonies**. Initially this provided them with mutual support, but as groups became permanent it allowed for specialization. Each cell in a permanent group does not need to perform all the functions necessary for life. Some can divert their energies to gather food, store it, defend the colony or attack its neighbours, reproduce, move the group around, or perform some other limited task. Groups of cells that learned to collaborate in this way began the long process of evolution into plants or animals.

When such a group had grown to a certain size the food needed by a closely-packed mass of cells could no longer reach all of them, and the ones at the centre could not dispose of their waste without poisoning their neighbours. The formation of internal spaces filled with fluid was the solution to the problem. Nutrients could pass through these spaces to reach all parts of the organism, and they were also used for the disposal of waste. Later the same channels acted as conduits for the chemical messengers that evolved to link and coordinate more complex functions as they developed. The human heart and circulation is an example of the most sophisticated of these systems in existence today.

Single-celled microbes did not disappear as the multiple-celled forms of life evolved. The tissues and liquid-filled spaces of the early animals and plants would have made ideal havens for the single-celled microbes that still surrounded them. On offer was the attraction of five-star hotel accommodation compared with the usual precarious existence of microbes that have to live rough. There can be no doubt that all primitive animals and plants were forced to develop defences designed to exclude microbial predators that would otherwise have invaded and destroyed them. This must have happened early in evolutionary history, and the human immune system shows how complex these countermeasures have become.

Microbes were and still are the most numerous living things on earth. Early in their existence they had to overcome a problem if their success was not to be their downfall. Smaller molecules taken from the environment were locked up inside the larger molecules of living cells and in their remains after death. If this had continued unchecked the environment would have been choked with the corpses of microbes, plants and animals, and at the same time exhausted of some of the components needed to make new living things. Fortunately some forms of life found that other microbes, plants and animals were good sources of food, particularly when they were dead and unable to defend themselves. These scavengers discovered how to break down and use the large molecules of their prey and they released the vital components back into the environment to be used again. Most of the microbes that exist today fall into this useful, indeed indispensable, category.

Some microbes find that life close to or even on their larger neighbours guarantees a regular supply of food, and provides them with shelter. These colonists do no harm so long as they confine their attentions to their hosts' waste and remain on their body surfaces. All animals and plants are **colonized** by a **normal flora** made up of microbes that remain harmless (as **commensals**) for so long as they obey these two rules. The microbes of the normal flora of some plants, insects and animals even pay for their keep by making certain molecules their hosts need but cannot produce for themselves.

1.2 HOW MICROBES ARE NAMED AND SEPARATED

Though Shakespeare wrote 'What's in a name . . . a rose by any other . . . would smell as sweet . . . ', scientific life is not so simple. To avoid confusion it is necessary to be precise when something new is described and a record is made. The science concerned with the naming of living things is called **taxonomy**. Only a little will be said about this now, but there is more detail in Appendix 1. The scientific names of living things are written in Latin, and are made up of two or more words. For example among some 300 types of oak the common European oak is *Quercus robur*, while the American white oak is *Quercus alba*. The evergreen holm oak is *Quercus ilex*. All oaks have botanical similarities so are allocated to the same genus with the first, generic name *Quercus*, but the different types are distinguished as different species by their second, specific names.

In the cases of animals and plants these distinctions are made by the use of features such as shape, size, colour or other anatomical or evolutionary criteria of an objective kind. Such rules are not easily applied to microbes. They are too small to be divided into genera and species by means of their size and shape. As determined with even the best optical (light) microscope these criteria permit only a superficial classification. For greater precision more reliable features must be used. As in other cases structure is the ideal criterion, but with microbes this must be examined with an electron microscope or chemically, at the level of individual molecules.

Before this was possible tests of a much more subjective kind were applied to separate microbes, and in particular bacteria, into genera and species. Many of these tests are still used in laboratories today. They often involve simple physiological functions that may be identified as present or absent, in a test-tube. For example, some kinds of bacteria can use the milk-sugar lactose for food, others cannot. The results of a test designed to detect this divides bacteria into 'lactose fermenters' and 'non-lactose fermenters', respectively.

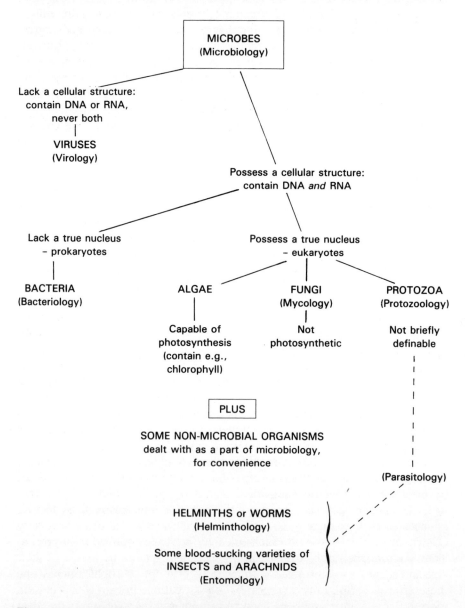

Figure 1.1 A classification of the microbes and an outline of the scope of microbiology. The names of the microbes are given in capital letters, and those of the associated sciences are in parentheses.

When the results of a number of tests of this sort are added to information about the size, shape and arrangement of bacteria they are progressively subdivided into smaller and smaller groups. Eventually they may be allocated to a single species.

The different forms of microscopic life are outlined in Figure 1.1, together with some larger parasites that are customarily included in the study of microbiology. Microbes are divided into bacteria, viruses, fungi, algae and protozoa. It is usually easy to distinguish between typical members of each of these five groups, but at the boundaries between them microbes are found that are not allocated so readily. For example, the parasite *Pneumocystis carinii* has at various times been classified as a protozoon or as a fungus, with recent emphasis in favour of the latter.

Microbes are, by definition, small. The unaided human eye cannot easily see anything less than one-tenth of a millimetre across. The smallest object that is visible under the best light microscope measures about a half of the wavelength of light, or about a quarter of a micrometre (μm). A μm was once called a micron. A thousand μm add up to one millimetre). In an ordinary optical microscope light shines through the object to be studied. For this to happen the object must be very thin and relatively transparent, as is the case with a microbe. Otherwise it has to be cut into thin slices. Objects to be examined are usually stained with dyes chosen to make them stand out from their surroundings.

To make tiny unstained fragments of jelly visible the optical tricks of **phase-contrast** or **dark-ground** microscopy are used. In fluorescent microscopy the object is illuminated with invisible ultra violet (uV) light after it has been stained with a dye that can convert uV into visible light. Those parts of the object that have taken up the dye are then seen to shine brightly against a dark background. This technique has been adapted to examine the detailed chemical structure of microscopic objects. It can be used to detect microbes by treating a smear or section of a tissue with an antibody to that microbe, after the antibody has been 'labelled' with one of these special dyes.

Nearly all the viruses measure less than 0.25 μm across, so were invisible until the electron microscope (EM) was invented. An EM employs a beam of electrons in place of light, and electromagnets instead of glass lenses. The result is a machine that is much larger and more expensive than an optical microscope, and about 200 times more powerful. The first virus was seen under an EM in 1939.

Most microbes are beneficial to the human race but there are potential adversaries in each of the groups into which they are divided (Figure 1.1). One of these groups, the algae, can be dismissed in a few words. When algae grow vigorously in water they produce various toxins. Humans may suffer from these when they come into contact with such water at work or in play (Elder *et al.*, 1993). The other groups and the larger parasites must be considered in rather more detail.

1.3 INFECTION

Microbes cease to be harmless if they, or harmful substances they make, penetrate into the tissues of a host. When an entry has been made microbes use the tissues that surround them for food and the harm this does is compounded by their toxic waste and by any offensive or defensive substances they produce. The study of the damage that results belongs to the science of pathology. The diagnosis and treatment of those who suffer as a result is the study of infectious disease. The microbes responsible for the damage are, or have become, **pathogens**. The study of the microbial pathogens that afflict different forms of life divides the science of microbiology into plant, veterinary and medical branches. Medical microbiologists study the pathogenic microbes that attack the human race. The infections they cause are the province of infectious disease (ID) physicians. The way in which microbes get about, and how to prevent this is a part of epidemiology, studied by epidemiologists. Because the three subjects are so closely connected there is a great deal of overlap between the territories of these specialists.

Because plants, animals and humans carry a normal microbial flora it does not follow that they suffer from continuous infections. For an **infection** to exist the host and the parasitic microbe must react together in some observable way. In a human being this reaction may be insignificant and only detected by a laboratory test. When this happens the infection is **subclinical**, **asymptomatic**, or **inapparent**. If pain, fever or other symptoms cause uneasiness or 'dis-ease' the individual suffers from a **clinical infection**.

Pathogenic and potentially pathogenic microbes do not have it all their own way. It has been noted that some microbes have developed countermeasures to the attacks of other microbes. When primitive microbes grouped themselves together to form larger organisms, these defences were developed further by specialized cells, dedicated to the task. The result is that the blood or equivalent liquid of even quite simple forms of life contain sophisticated mechanisms designed to attack and destroy microbes or other foreign materials that invade them. In this context foreign materials include any of the organism's own cells that have become abnormal or have died.

These mechanisms have been studied in some detail. Some of them operate in a series of stages so that a tiny reaction at a first stage triggers a larger reaction at a second, and so on. In this way a small biological event can be amplified to produce a quantity of an active end-product. This may be designed to bring considerable destructive power to bear on an invader. If this cannot resist the attack it is destroyed. These 'cascade' reactions are joined by other defences based on individual cells and by some that are designed to protect whole organs. Taken together these defences are the **immune mechanisms** that will be described in Chapter 2. Sometimes these mechanisms backfire, that is, if they are over-stimulated defensive chemicals may be produced in excess, in the wrong place. Host tissues are then attacked whether or not microbes are present. This happens in some chronic diseases and as a terminal event in severe infections.

Infections are unique among diseases because they involve an interplay between two distinct living systems. Each is capable of infinite variation. On the microbial side the most important variation is the capacity to produce pathology, or the **pathogenic potential** of the microbe. This can vary from nil for microbes that are incapable of causing an infection (a **non-pathogen**), through microbes that are weakly pathogenic (**opportunistic** or **occasional** pathogens), to microbes that are **fully** or **highly pathogenic**. On the human side there are wide variations in the effectiveness of the defence mechanisms. Any or all of them may be weakened or absent. A reduction in efficiency (**immunodepression**) may be temporary or permanent, accidental or intentional, congenital or acquired. Normal immunity or **immunocompetence** then gives way to temporary **immunocompromise**, permanent **immunodeficiency** or deliberate **immunosuppression**. When a microbe attacks someone the outcome depends on the point of balance struck between the two sides. Because each side varies so widely the outcome is difficult to predict. Of three people exposed to identical risks one may suffer a severe or fatal infection, the second a mild illness, while the third escapes entirely.

Because infections are caused by interactions between two living systems the study of microbiology on its own gives a very blinkered view of the subject. To complete the picture the diseases that result and their epidemiologies must be included. To isolate one of these topics from the others is to hope that a stool with less than three legs will stand.

1.4 HISTORY

The origin of the study of infection is lost in antiquity. The separation of diseases into those that are infectious and those that are not is still incomplete. In the public mind the subject is surrounded by some confusion, born of ignorance. Epidemiology is in the same position. Ignorance of microbiology is also common but by comparison this subject has a short and comparatively clear-cut history. When microbiology emerged as a new science towards the end of the nineteenth century the study of infection and its epidemiology began to develop on a scientific basis. It became possible to make orderly advances in the diagnosis, treatment and prevention of infection.

Before this time accurately targeted advances had been rare, but some are worthy of note. One was the introduction of vaccination as a safe way of preventing smallpox (Box 1.1). In 1796 this was put on a firm foundation

---

Box 1.1 Variolation against smallpox

Lady Mary Wortley Montague was married to a dull diplomat. Her sharp wit matched a sharp mind and she lived such a full life that her diaries were destroyed after her death. At the age of 28 she went to Constantinople with her Ambassador husband. In 1717 she wrote home that ' . . . the smallpox, so fatal and general among us, is here entirely harmless by the invention of ingrafting . . . old women perform the operation every autumn . . . they make parties for the purpose and the old woman comes with a nutshell full of the matter of the best sort of smallpox and a large needle (and), gives . . . a scratch. . . .

This practice, variolation, had been common in China and India since ancient times. The deliberate inoculation of a mild form of smallpox had a mortality of perhaps 2% compared with about 30% for the usual infection. Back home in England Lady Mary campaigned for variolation. She arranged for seven criminals to be offered variolation in place of the gallows. They accepted, survived, and were freed! Variolation was made illegal in Britain in 1840, by which time Jennerian vaccination with cowpox had been established for over 40 years.

by the Englishman Edward Jenner. In 1847 the Hungarian Ignaz Semmelweis worked out how to prevent the spread of fatal childbed fever or puerperal sepsis, a life-saving discovery unhappily ignored at the time.

In the midst of a cholera epidemic in 1855 John Snow removed the handle of a pump in London, England. He had studied the way cholera spread and concluded that the pump was connected to a contaminated well. He proved his point when his action stopped the local outbreak, and he showed for the first time that water plays a part in the epidemiology of the disease. For the human race the beginning of the twentieth century marked a turning-point in the conquest of infection. The credit for this is shared between the new scientific approach made possible by microbiology, and a general improvement in nutrition and housing. It is unfortunate that much of the population of the world has yet to feel the benefit.

The existence of forms of life invisible to the naked eye was the discovery of a Dutch amateur lens maker, Anthonie van Leeuwenhoek (1632–1723). He constructed a powerful magnifying glass through which he could see microbes. The 'animalcules' he described excited philosophical curiosity, but their scientific significance was not recognized for another 200 years. In the interval the compound optical microscope was developed. When this had reached a sufficient state of development the stage was set for the key event in the history of microbiology. This was the invention of methods for the cultivation of microbes, so they could be studied in the laboratory.

Box 1.2 Louis Pasteur (1822–1895)

Louis Pasteur was the son of a sergeant-major who served under Napoleon in the French army during the Peninsular War. He started life as an organic chemist, but soon became the first microbiologist. An industrialist with a problem asked for his help. He produced alcohol from beetroot juice, and something had gone wrong. Pasteur found that some of the vats contained bacteria that made lactic acid from the beet sugar instead of yeasts that turned it into alcohol. He invented pasteurization and controlled the problem.

Later he disposed of the theory that life could be generated spontaneously and discovered the cause of a disease of silkworms that had ravaged the French silk industry. He also produced vaccines against fowl cholera, anthrax and swine erysipelas, and one against human rabies. The Pasteur Institute in Paris is his memorial.

The Frenchman Louis Pasteur (1822–1895) began to develop these techniques as a part of his work on fermentation in the production of alcohol, vinegar and other products (Box 1.2). Many of his methods are still used today. Pasteur published the first description of the activity of a microbe in 1857, and although most of his achievements were in the industrial and veterinary fields, he is honoured as the father of medical as well as of general microbiology. Where Pasteur led, others followed. Robert Koch (Box 1.3) and a growing band of other disciples expanded the science, and identified an increasing number of individual microbes. The causes of many important infections were identified for the first time between 1875 and 1900.

Box 1.3 Robert Koch (1843–1910)

Robert Koch was one of 13 children, the son of a mine overseer. He studied medicine in Berlin and Hamburg and served in the army in the Franco-Prussian War (1870–71). He was then appointed medical officer to the small Prussian town of Wollenstein. His wife gave him a microscope for his 29th birthday and he set up a laboratory at the back of his consulting room. Here he began his researches on anthrax (there were no health and safety inspectors then!) and invented techniques that were central to the development of microbiology.

He then moved to a better job in Berlin and in 1881 presented his work at the Seventh International Medical Congress held at King's College, London. Louis Pasteur and Joseph Lister were also present. Because of the emnity generated by the war, Lister had to entertain the French and German delegations on different days. Pasteur was barely able to speak to Koch, so the two leading men in the same field of science failed to establish a working relationship. In 1882 Koch discovered the tubercle bacillus. In 1905 he was awarded the Nobel Prize.

It soon became clear that microbes could be divided into groups that possessed quite different characteristics, and microbiology began to break up into sub-specialities along these lines. The study of bacteria was called **bacteriology**, and this branch developed particularly rapidly in the early days. Another branch (**mycology**) concerned itself with fungi. Many fungi develop structures that are visible to the unaided eye so scientific mycology was founded before bacteriology. The subject did not attract much attention and the priority was swept away by Pasteur's highly pub-licized work. In 1901 yellow fever was the first human disease proved to be caused by a virus. Nearly all viruses are too small to be seen even with the best ordinary microscope, but they are included among the microbes. Because of their small size and their requirement for living plant, animal or microbial cells inside which to grow, initial progress in the science of **virology** was slow. Viruses lack so many of the characteristics of living organisms that they are not really alive. In the last resort the decision turns on how 'life' is defined.

To a great extent the study of protozoa (**protozoology**) depends on seeing and describing these microbes rather than on their cultivation. Most of them are rather small and not easily distinguished from the cells of their hosts among (or in) which they live. They were not reliably identified in specimens from humans until microscopic techniques improved towards the end of the nine-teenth century, so the science of protozoology developed simultaneously with bacteriology. The protozoon cause of malaria was first seen in the blood of a patient in 1880, though its complete life-cycle was not worked out until 1948. In 1886 another protozoon, the amoeba, was shown to be the cause of one kind of dysentery.

The study of worms (called **helminthology**) is often added to protozoology under the broad title **parasitology**. This may be extended to include noxious insects and similar pests that are properly a part of the science of **entomology**. The worms that invade humans are too big to be microbes as their adult forms at least are clearly visible to the naked eye. The existence of human worm infestations has been recognized for centuries, though the science did not develop until more recently, again because efficient microscopes were needed to identify the extraordinarily complex life-cycles of many of these parasites.

When infections are considered thought must be given to the hosts that are attacked as well as to the microbes responsible. The ability of a human host to resist or overcome an invader is measured by his or her immunity. The mechanisms of immunity are studied in the science of **immunology**. This science was also founded in the last quarter of the nineteenth century, but at first it developed slowly. The pace quickened in the 1940s as the result of the fundamental discovery that lymphocytes have a central role in the immune process. Advances since then have increased the understanding of infection and have stimulated developments in transplantation and the treatment of cancer.

1.5 ABOUT THIS BOOK

Newcomers to the study of the microbiology of infection are confronted by many new words and ideas. Nature frustrates teachers who would like to present these in an orderly fashion. Although microbes can be described in a logical progression the infections they cause overlap so much that simplification is impossible. A single microbial species may be involved in several distinct diseases, of different organs. For example *Staphylococcus aureus* may cause boils, styes, carbuncles, abscesses, wound infections, osteomyelitis, urinary tract infections, endocarditis, septicaemia and food poisoning, and the list of staphylococcal infections is still incomplete. Equally, one kind of infection can be caused by a number of different microbes. Cases of meningitis indistinguishable at the bedside have many causes. Among the bacteria the most common are *Neisseria meningitidis*, *Haemophilus influenzae* and *Streptococcus pneumoniae*.

The microbiology and epidemiology of infection may be presented in two different ways. The first approach describes the microbes individually. The infections to which each gives rise are introduced in the process. The second is to present each type of infection and within this format to give details of the microbes responsible. Because of the multiple overlaps just noted either approach tends to be repetitious, and neither is entirely satisfactory.

In an attempt to minimize these difficulties this chapter, Part One of the book, is provided to link the other eight parts together. Part Two describes the general qualities of the human and microbial antagonists that combine to cause, or prevent, infections. Parts Three through Seven introduce individual bacteria, viruses, fungi and other parasites that attack humans and then deals with some of the more common kinds of infection. Part Eight outlines the methods used to diagnose and treat infections, and Part Nine is concerned with their prevention.

FURTHER READING

1. Elder, G.H., Hunter, P.R. and Codd, G.A. (1993) Hazardous freshwater cyanobacteria (blue-green algae), *Lancet*, **341**, 1519–20.
2. See Appendix B.

Human and Microbial Interaction

PART 2

Human immunity $\boxed{2}$

2.1 INTRODUCTION

In Chapter 1 the subject of immunity was introduced in an evolutionary context. Here we are concerned with how it works. The immune system operates to keep the internal environment of the body free of foreign material. Foreign material ranges from some part of the body that has died or has strayed sufficiently from normality to be recognized as 'not self' to something more obviously foreign like a thorn in the finger or a living invader. The strength of the body's reaction varies according to the degree of 'foreignness' it detects. For example, an artificial hipjoint is clearly foreign, but it is carefully constructed to excite the least possible immunological reaction. A transplanted kidney is also foreign, but in this case it is necessary to use drugs to suppress the recipient's immune system to prevent the reaction that would otherwise reject and destroy it. Microbes are even more distinctly foreign so they excite very vigorous and potentially destructive immunological responses. If a microbe is to cause an infection of any consequence it must avoid these reactions, at least for a time. Some of the strategies developed by microbes to do this are described in Chapter 3.

In a book about infection we are concerned with the reaction of the body to microbes or other living things that may invade it. There are other important activities of the immune system that are neglected here, but it should be noted that immunity does not only function to control infection. If immunity fails the result may be the development of an infection or a cancer or an auto-immune disease, one at a time or all at once. The immune system is made up of many parts. These may act singly, but more often they operate in concert in ways that are not yet completely understood. As each section is described it may not be clear how the parts fit together. To overcome this Figure 2.1 and Tables 2.1 and 2.2 provide a summary of the subject.

2.2 THE FIRST LINE OF DEFENCE

The surface of the body is the first and most important line of defence against invasion from the outside. It is made up of a variety of membranes and is much more extensive and complex than is conventionally supposed. It consists of a comparatively small stretch of skin to which areas equivalent to several tennis-courts are added when the surface linings of the gastrointestinal

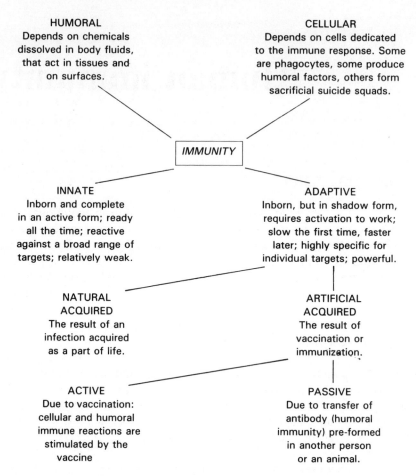

Figure 2.1 The compartments of the immune system, and how they are related.

canal, the lungs and the urogenital tract are included. In females even the peritoneal cavity is a part of the surface of the body (it is open to the outside through the fallopian tubes). A little thought will show that none of the organs concerned with these arrangements could carry out their physiological functions if this were not so. In practice most of these very large surfaces are intricately and conveniently folded away, out of sight.

From the microbiological point of view the surfaces of the body are of two kinds (Figure 2.2). Some are permanently colonized by a large population of microbes (the **colonized surfaces**), others are kept substantially free of them (the **'privileged' surfaces**). The main colonized surfaces are the skin and its appendages, the gastrointestinal canal and the vagina in the female. The privileged surfaces are provided with mechanisms that exclude microbes. The lungs, urogenital tracts of both sexes other than the vagina, the pancreas and the biliary system of the liver are the major privileged surfaces.

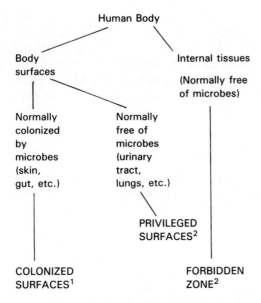

Figure 2.2 The distribution of the normal human microbial flora on the surfaces of the body.

2.2.1 Colonized surfaces

Very large numbers of microbes inhabit the colonized surfaces. Each individual carries at least 10 000 times more bacteria than there are people on earth. This does not mean that our bodies are a free-for-all microbiological playground. Sweat and sebaceous secretions make the skin acid so any microbes that survive there must be able to tolerate this. Other variations depend on how dry or wet a surface is, and how much oxygen is available. The environment at the surface of the mucous membranes of the gastro-intestinal canal and vagina are controlled by these factors. Acid is again important. The strong hydrochloric acid secreted by the stomach kills or disables most of the microbes present in what we eat and drink before they can reach the intestine and multiply there. The healthy pre-menopausal adult vagina is populated by the kinds of commensal microbes that can survive in the acid they themselves make from the glycogen thoughtfully provided by their host. Individuals who have never met carry similar numbers and types of microbes because the surfaces on which these live are also similar.

A well-established normal (or commensal) flora is itself a barrier to potential colonists. Microbial species of all kinds exist in a perpetual state of war. They possess weapons of offence and defence. The large population of permanent residents that occupy different parts of the body are well placed to repel smaller numbers of invaders. Even if they are temporarily over-whelmed by a large horde the invasion fails if the conditions on the surface are unsuitable for the newcomers. However, if the surface is damaged, the environment is changed so otherwise unsuitable invaders can more easily set up house and make themselves at home. The same thing may happen when antimicrobial drugs are taken, or antiseptics are applied to the skin or mucous membranes. The usual residents suffer heavy losses and this weakens their opposition to new arrivals. The new colonists that settle in may not be as friendly as the old ones.

2.2.2 Privileged surfaces

At some point each privileged surface is in structural continuity with another that is heavily colonized. Compare for example the enormous number of microbes present in the throat, and the relative lack of them in the trachea, yet the two sites are only millimetres apart. How are these separations achieved? The mechanisms responsible are not completely understood. Billions of bacteria in the faeces of a baby girl are spread onto her perineum every time she defecates, yet they very rarely penetrate to cause infections in her urinary tract. The infant female urethra achieves this exclusion though it is much shorter than its male counterpart. It is washed through each time she urinates, and her peri-urethral glands secrete antimicrobial substances. These defences may seem weak yet they exclude the inhabitants of the perineal skin and a faecal challenge that is repeated at least daily for months on end.

Concentrated bile has a powerful antimicrobial effect, so this is a good deterrent to the spread of microbes into the biliary tract and the liver. It is also easy to understand some of the mechanisms that keep the respiratory tract clear of microbes. The most obvious are the violent expiratory explosions of sneezing and coughing. When these are triggered by irritation of the mucous membranes of the nose, larynx or trachea, any larger particles present are expelled.

The machinery that keeps the normal flora of the pharynx out of the neighbouring trachea and the rest of the lower respiratory tract is more complex. The same mechanisms operate to remove the much smaller number of microbes present in the considerable volume of air inhaled every day. Even apparently clean air contains surprisingly large numbers of very tiny particles. The myriads of dancing specks visible in a shaft of sunlight in an otherwise darkened room shows how many there are. Most are inert fragments shed from the surfaces of humans, animals and plants, or are particles of other debris. In occupied buildings a large proportion of the particles are the skin squames people shed in large numbers every day, plus tiny fibres rubbed from their clothing. Very few of the particles consist of microbes, or even carry them. If all the rubbish in inspired air was retained in the lungs the bronchioles and alveoli would soon be clogged with it. Respiratory insufficiency and infections would follow and respiratory function would soon cease.

The most important mechanism that prevents this is illustrated by the observation that if a car fails to take a corner because it is going too fast, it slides off the road. Before the air we breathe reaches the alveoli in the lungs it passes through tubes (the trachea, bronchi and bronchioles) that constantly change their direction. As the air travels fast round quite sharp corners the particles in it are thrown against the walls of the tubes. These are lined with sticky mucus so the particles do not bounce off, but are trapped. The mucus contains natural antimicrobial substances and is kept in constant motion, wafted along by tiny hair-like cilia. These are borne on the surfaces of the cells that line the naso-pharynx, trachea and bronchi. From the lungs the mucus moves upwards as if on an escalator, eventually to reach the larynx and then the pharynx. It is then swallowed or expectorated, together with the particles it has collected. Any particles that manage to avoid the mucus reach the alveoli where they encounter other defenders, the amoeba-like macrophages. These phagocytic cells engulf any particles that have avoided the mucus, and remove them.

All these mechanisms operate against particles that contain microbes as well as those that do not, so the silting-up of the lungs and chest infections are prevented at the same time. When the mucus escalator does not work chest infections are more common. The cause of the fault may be congenital, as happens in cystic fibrosis, or the machinery may be damaged by smoking or other inhaled toxins, or by such infections as whooping cough or influenza.

The 'antimicrobial substances' present in the mucus produced in the lungs are soluble chemicals that also appear in secretions on other surfaces of the body. At their simplest they are the acids that may kill microbes, or at least stop them from multiplying. The most complex are special forms of the antibodies to be described later. In between are a variety of substances toxic to microbes. One of these is **lysozyme**, a chemical found in tears and other secretions. This dissolves the cell walls of some bacteria, so they break open and die.

Much stress has been laid on these front-line defences not only because they are of considerable (and widely underestimated) importance, but also because they are of special relevance to health professionals. The mechanisms that have been discussed are easily damaged or inactivated by the techniques used by doctors, nurses and others in the diagnosis, treatment and care of their patients. It is surprisingly difficult to avoid doing at least some harm when attempting to do good. The deliberate use of drugs to suppress immunity and the potentially deleterious side-effects of antimicrobials and antiseptics have already been mentioned. Whenever defences at the body surface are inactivated or breached infections are more likely, or in the case of antimicrobial drugs an existing infection may be replaced by another that is worse than the first. These problems will be dealt with when infection is discussed in Chapter 4.

2.3 THE SECOND LINE OF DEFENCE

The defences that have been described so far are parts of the **innate** system of immunity. This is present at birth, or develops soon afterwards, and it operates indiscriminately against any material that it recognizes as foreign. Its counterpart is **adaptive immunity** (Table 2.1). Adaptive immunity is responsible for highly specific actions that are tailored precisely to attack

Table 2.1 An outline of the compartments of the immune system

	Innate	Adaptive
Cellular	Body surfaces Specialized cells (Table 2.2)	Specialized cells (Table 2.2)
Humoral	Complement Blood clotting Lysozyme, etc. Acute-phase proteins Interferons	Antibody

a particular part of a particular invader and, ideally, nothing else. The adaptive mechanisms are much more powerful than the broad-based, ever-ready innate system, but they take time to develop their full potential against a new invader. The two systems are at their most effective when they work together. Both of them operate throughout the tissues of a healthy body. In normal circumstances an invader that breaches a surface immediately encounters a second line of defences. Some of the defenders it meets are chemicals, dissolved in body fluids. For centuries these fluids were the 'humours' (Box 2.1) so these chemicals were called 'humoral factors' that contribute to **humoral immunity**. They are to be distinguished from defences that depend on the whole specialized cells responsible for **cellular immunity** (Table 2.2).

Box 2.1 The humours

The ancient philosophers Hippocrates, Aristotle and Galen considered that the body contained four kinds of fluid: blood, phlegm, yellow bile or choler, and black bile or melancholy. These were the cardinal humours, and an individual was sanguine, phlegmatic, choleric or melancholic according to which humour was dominant. Disease was accompanied by a disorder of the humours, and treatment was designed to correct the imbalance, for example by purging or blood-letting.

In Europe this view of physiology persisted well into the 18th century.

2.3.1 Cellular immunity – phagocytes

The cells that are included as parts of the cellular system of immunity have no other functions than immunological activity. These are outlined in Table 2.2. Enormous numbers of these cells are distributed throughout the body. The first of their functions to be described was **phagocytosis**, the eating-up of foreign material. Elie Metchnikoff discovered this phenomenon well over a century ago (Box 2.2). Human phagocytes are of two basic kinds, the **neutrophil polymorphonuclear leucocytes** (PML, polymorphs, polys or pus cells for short) and the **monocytes** and **macrophages** that are the same cells found in different places.

The phagocytes are part of the innate system of immunity. On their own they are not very effective. Foreign particles are made more attractive by

Table 2.2 Contributors to the cellular category of immunity, with the functions of the cells concerned

CELL TYPES	FUNCTIONS	IMMUNE SYSTEM
Polymorphonuclear leucocytes		
neutrophil	Phagocytic cells, originally called microphages (compare macrophages, below). Particularly active against pus-producing microbes.	
eosinophil	These attack larger parasites, e.g., worms.	
basophil	These have many functions including the initiation of the inflammatory response and allergic reactions. When found in the tissues outside blood vessels they are called **mast cells**.	All these cells are parts of the innate system of immunity. Their actions are enhanced by certain of the innate humoral factors, and are very powerfully enhanced by the adaptive system of immunity.
Monocytes and Macrophages	These similar, larger phagocytic cells are found in the blood and tissues, respectively. They are specially active against intracellular microbes.	
Lymphocytes 'NK' cells	Natural killer cells: attack cells whose surfaces are altered because they contain microbes, have become malignant or are foreign (transplants).	
'B' cells	Antigen recognition and conversion into cells that produce humoral antibodies.	These cells are central to the adaptive system of immunity.
'T' cells	Antigen recognition and activation in cell-mediated immunity, the adaptive equivalent of NK cells.	

Box 2.2 Phagocytosis.

Elie Metchnikoff (1845–1916) was the youngest son of a Russian landowner who had squandered the family fortune. His mother would not allow him to do medicine, so he studied zoology instead. As an adult he developed a depressive illness, about which he kept notes. He tried to commit suicide first by infecting himself with relapsing fever (he recovered) and then with opium (he survived). At 30 he married a girl in her early teens, and received an inheritance from her parents. They rented an apartment overlooking the Straits of Messina in the Mediterranean. In 1884 when studying the transparent larvae of starfish with his microscope, he noticed that certain cells moved around inside them. He took some rose thorns and pushed them into the larvae. In a few hours they were surrounded by the motile cells. Next he introduced microbes, with the same result. He formed the theory that these cells, phagocytes as he called them, were there to protect the larvae.

Surprisingly his theory met with initial indifference or even hostility. Robert Koch rejected it out of hand. Louis Pasteur had more insight, and Metchnikoff worked in Paris for the rest of his life. He was awarded a Nobel Prize in 1908.

the addition of humoral 'flavour-enhancers' that make the phagocytes much more voracious. These enhancers may be produced by the innate, or better still, by the adaptive system of immunity. Phagocytosis begins with adhesion between the phagocyte and its intended victim. Without this essential first step the target would float away each time the phagocyte tried to seize it. (Try biting an apple floating in a bowl of water without using the hands!) Engulfment follows and the victim is enclosed in a vacuole in the cytoplasm of the phagocyte. The process is completed when the phagocyte fills this vacuole with a highly toxic mixture that kills and may also dissolve the intruder.

2.3.2 Cellular immunity – lymphocytes

The 'B' and 'T' lymphocytes are highly specialized cells that belong to the adaptive immune system (Table 2.2). They are small lymphocytes that have been educated in one of two special schools. One of these is sited in the thymus gland, hence the 'T' label for the graduates that emerge. The other school was first detected in a structure in chickens called the bursa of Fabricius. Its products are still called 'B' cells, though this organ has no direct counterpart in the human body.

'B' cells are responsible for the production of antibodies. Antibodies belong to a family of similar proteins called **immunoglobulins** (Box 29.4). Each member of the family has a slightly different set of properties and functions but their similarities outweigh their differences. At a simple level an immunoglobulin molecule can be thought of as a tiny Y-shaped string of protein. At the top of each arm of the 'Y' are the parts that recognize and adhere to antigens, in a highly specific way. An antigen is any molecule of a rather large size. In the present context it is any foreign molecule that is identical to the one that originally called a particular antibody into existence. Antigens may float about as single molecules, or they may be part of a larger invader, or be sited on or in any of the cells of the body. Antigens and antibodies belong to each other individually and specifically just as a lock is opened by a key of only one shape. At the lower end of the 'Y' is a region that is activated when an antigen unites at the other end. When activated this region sets in train various reactions. Antibodies exist in solution in body fluids so are part of the humoral system to be described more fully later.

The mass of 'B' lymphocytes that circulate through the body are provided with an astonishing array of receptors for foreign molecules. These are required to recognize and respond to a huge range of potential antigens, yet each lympocyte only carries a receptor for one of them. Although there are a great many 'B' cells, there are so many possible antigens that only a very few lymphocytes can be available to detect and react to any antigen that has not been met before.

Each 'B' lymphocyte waits for the chance that it might meet and unite with its own particular antigenic counterpart. When it does so this single lymphocyte cannot produce a useful amount of antibody. When it meets its own antigen the first thing that happens is that the lymphocyte is stimulated to multiply to form a large family of identical lympocytes. When there are enough they form a factory that produces quantities of antibody. This antibody is entirely specific for, and can only unite with, antigens identical to the one that started the process. The antigenic lock and the now multiple antibody keys match perfectly. When a new antigen is met for the first time, perhaps as a part of a microbe not previously encountered, antibody production takes some time to get started. After seven to ten days there may be enough of it to be detected in the laboratory. Once the factory is set up it continues in existence, perhaps for many years after its initial task is complete. If the same antigen (or the same microbe) reappears later the factory is available to swing back into production much more quickly, this time in a matter of hours. This is how the slow initial or **primary response** contrasts with the much more rapid and productive **secondary** one.

'T' lymphocytes are selected, activated and multiply in a way that is broadly comparable to their 'B' cousins. However the end-product, **cell-mediated immunity**, is quite different from the humoral antibody just described. The progeny of a 'T' lymphocyte that has been activated by its own antigen forms a group of cells that individually attack any targets that bear the same antigen. The targets are usually other host cells that have on their surfaces tell-tale antigens that indicate that all is not well inside them. The antigens are identical with those that activated the original 'T' cell and so started the whole process. They provide a signal that indicates that the cells to be attacked have developed undesirable characteristics so need to be destroyed. These signal molecules may be by-products of microbial multiplication going on inside a cell, that would otherwise be hidden from the immune system. Other signal molecules may indicate that the cell concerned has escaped from control to become 'malignant', so if not destroyed might multiply to form a cancer. Sometimes, less helpfully, the signal molecules tell 'T' lymphocytes that a surgeon has slyly substituted a foreign organ for the one their owner was born with, but which became defective.

This last example prompts the question, why do these remarkable 'B' and 'T' cells not act in response to the many antigens that surround them in the body of which they are a part? The answer is, sometimes they do. The result is an occasional immunological or auto-immune disease. The more usual negative answer depends on the fact that fetal immunological reactivity develops late in pregnancy, or shortly after birth. During the early part of this period any lymphocytes that might react with 'self' tissues are destroyed. If all goes well those that remain are the ones that can only react with dangerous foreign or 'non-self' antigens that the individual may meet and need to attack later in life.

In one sense this late development is just as well, as an infant might otherwise begin to develop humoral or cellular immune reactions against its mother's tissues while still inside her. In another sense it is a pity, because for some months after birth infants are more susceptible to infections. This inadequacy is partly covered by the ante-natal passage of humoral antibodies through the placenta from their mothers (passive immunity, Figure 2.1). These antibodies last for six months or so after birth, usually long enough for the infant to complete its own defences. Another question is why a mother does not reject her fetus shortly after it is conceived? A fetus is a 'transplant' made up of tissue one-half of which is potentially foreign, or more than half in the case of *in vitro* fertilization with donor eggs. Most of the answer has no connection with infection, but pregnancy is in fact accompanied by a minor degree of immunosuppression. The result is an increased susceptibility to infections like listeriosis (Chapter 7).

The activities of 'B' and 'T' cells and of adaptive immunity account for the common observation that it is unusual to suffer from second attacks of many of the common infections. The immunity depends on a memory of the first attack. This lies in the residue of the production lines for humoral and cell-mediated immunity that were set up in the first place. The immunity is reinforced if or when the same microbe or antigen is met with again later in life, or, as happens with chickenpox for example, the microbe is not completely eliminated after the infection. In cases such as this the infecting organism establishes a permanent, shadowy existence somewhere in the body that may provide a continuous stimulus to the immune system. This freedom from repeated attacks of an infection is due to **natural acquired immunity** (Figure 2.1).

The other kinds of lymphocytes in the armoury of cellular immunity are the **natural killer** (NK) cells (Table 2.2). These sacrifice themselves for the benefit of the community in a fashion similar to activated 'T' lymphocytes, but in this case they seem to react to a number of different trigger molecules on cell surfaces. Unlike 'T' cells they are not activated by, nor do they multiply after, initial contact with an antigen. They are ready for instant action against any cell that bears a signal they recognize as foreign, so belong to the non-specific innate system of immunity. They are particularly useful against cells that have been invaded by viruses.

2.3.3 Cellular immunity – other cells

The roll-call of cellular immunity includes three other types of cell. These are the **eosinophil** and **basophil** polymorphonuclear leucocytes (eosinophils and basophils for short) and **mast cells**. Basophils and mast cells are essentially the same, but are stationed in the blood and tissues, respectively. Eosinophils are specialized to attack invaders, such as worms, that are too big for phagocytes to handle. In this task they are helped by complement (described below), and antibody. Basophils and mast cells are important accessory factors in the production of an inflammatory response in which they also act with complement together with phagocytes. They have an additional, less attractive function. Together with a special class of antibody (immunoglobulin 'E', IgE, Box 29.4) they are the cause of allergic reactions such as asthma and hay fever that may afflict up to 10% of the population.

2.3.4 Humoral immunity – innate

Two innate humoral systems are of great importance in the fight against invaders. Both are cascades, so consist of a number of chemicals that react in series, with a small change at an initial stage that is amplified in the next, and so on. The first of these is the mechanism responsible for blood-clotting. Although usually thought of in connection with the prevention of bleeding, the conversion of soluble fibrinogen into a clot of insoluble fibrin in the neighbourhood of invading microbes serves to immobilize them and so limit their spread.

The **complement system** is responsible for the second cascade reaction. Complement is intimately concerned in the production of **inflammation**, so is central to the process of infection. An infection, a boil or a sty for example, is accompanied by redness, heat, swelling, pain and loss of function, the five cardinal signs of inflammation. The complement cascade may be switched on in its innate immune mode by molecules that are parts of the surfaces of invading microbes or by acute phase proteins that have been activated by them. In its adaptive mode the trigger is an activated antibody molecule. In either case a number of new substances are formed that initiate further reactions. Put together these have powerful and far-reaching effects.

The first of these reactions is to stimulate phagocytes and other special cells to release yet another array of active chemicals. These cause blood vessels to dilate in the neighbourhood of complement activation. The result is that the area becomes red and hot. The permeability of the vessels is also increased. Blood plasma escapes from the dilated vessels to bring additional complement and other active compounds to the right place to continue and intensify the attack on the invader. This causes the part to swell, become painful and more difficult to use, to complete the cardinal signs of inflammation.

Another effect of these substances is to cause **chemotaxis**, a process by which polymorphonuclear leucocytes are activated and positively attracted to the place where they are needed to carry out their phagocytic duties. Other products of complement activation coat the outside of microbes so that they become sticky and adhere to phagocytes as an important preliminary to their engulfment and destruction. Yet another product is a mechanism which can punch holes through the membranes that form the boundaries of cells. This allows their contents to leak out, so they die. This mechanism is particularly active against 'foreign' cell membranes such as those found in microbes.

Some of the remaining humoral defences found inside the body are the same as those that may be secreted onto its surface. For example, the lysozyme present in tears is also widely distributed in the tissues. The **interferons** are another group of substances found in the tissues whose functions include an ability to cooperate with NK lymphocytes to slow down the rate at which viruses spread from one cell to another. Interferons are produced by cells in response to a variety of stimuli, including an attack by viruses. Other chemicals, collectively known as **acute-phase proteins**, are produced in large quantities in the early stages of many infections. These act against microbes in a non-specific way to activate complement and stimulate phagocytosis. They are much less good at this than antibodies but they are available to tackle a new infection in its early stages, long before antibodies are ready. One of them, called C-reactive protein, is easily detected in the laboratory. The level of this protein in the serum of a patient may be measured to help to distinguish between an illness due to an infection and one due to some other cause. In the former case the patient might benefit from an antimicrobial drug, while in the latter another diagnosis must be sought, and antimicrobial treatment may not be necessary.

2.3.5 Humoral immunity – adaptive

Antibody is produced by 'B' cells that have been stimulated by an antigen, as has been described. Initially this is a cellular activity, but the antibodies that are formed are part of the humoral compartment of immunity. The remarkable specificity of the relationship between an antigen and its antibody has been stressed. An antibody that is produced following an injection of tetanus toxoid is of no use against diphtheria toxin, and the reverse is also true. Antibodies are among the most highly developed manifestation of immunity. They are at their most effective when they act in conjunction with other components of the immune system.

The outcome of the union between antibody and antigen depends on the size of the structure to which the antigen is attached. A single antigenic molecule may function on its own as an offensive entity. This is the case with the molecules of diphtheria or tetanus toxin. Alternatively an antigen may be attached to a much larger invader, such as a bacterium or a worm. When an independent antigen of the smallest size unites with its particular antibody, the new structure differs markedly from either of its components in size and in other ways. The most obvious effect of the change is that the antigen is likely to lose one or more of the properties it previously possessed. For example, molecules of diphtheria or tetanus toxin united with their antibodies are no longer toxic. This is how the antitoxins produced in response to diphtheria and tetanus vaccines prevent diphtheria and tetanus.

The same thing happens with somewhat larger invaders such as viruses. If a virus enters the body of a host who has met the same microbe previously, it is coated with the antibody that is already present. The result is that the virus loses a function critical to its survival. This is the ability to adhere to the outside of the cell it must enter in order to multiply. When this happens an infection is snuffed out before it can begin.

In these examples toxin molecules and virus particles have lost their capacity to cause infections so they have been neutralized by the antibody. This is the end of the ability of antibody to act by itself. Compared with a molecule of antibody or even a virus, most bacteria are enormous structures. A bacterium is not seriously inconvenienced if a few, or even a number of molecules of antibody adhere to antigens on its surface. Other members of the immune system now join the fray. Antibody that has been activated by adherence to its antigen (in this case part of the surface of a bacterium) initiates the complement cascade and stimulates phagocytosis, with great efficiency. When they act together these three components of the immune system are a match for most microbes. Worms are even larger than bacteria. The special destructive power of eosinophils are deployed against them, again with enhancement when assisted by antibodies.

The effects described are beneficial to the host, and indeed benefit is the most common outcome of the action of an antibody. As usual, there is a down-side to the story. If for some reason large quantities of antigen and antibody are present simultaneously, particularly in the blood, large sticky 'immune complexes' form. These particles might cause problems even if they were inert, but the antibody components of the complexes are activated, so the complement cascade is switched on and phagocytes are stimulated. This may start an inappropriate and often widespread inflammatory reaction, much of it in the wrong place. This can cause severe damage to the host in the form of 'immune complex disease'. An example is the nephritis that may follow a streptococcal infection of the throat.

Antibodies may be produced that are able to react with normal components of the body. This may happen when the control mechanisms that should prevent it have failed to operate, or if lymphocytes are stimulated by an infection with a microbe that carries antigens similar to some of human origin. The antibodies that are produced are **autoantibodies**, and the result is auto-immune disease.

Another more common example of an undesirable action of antibody is the production of the symptoms of allergy. Some individuals may react immunologically to certain pollens. The type of antibody they produce may be an immunoglobulin of the IgE type (Box 29.4). This has a particular affinity for mast cells. The next time the individual meets the same pollen antigen, this combines with IgE, and mast cells are stimulated to release histamine-like compounds that cause the running eyes and nose, and the sneezing. Other manifestations of allergy are asthma and eczema.

The most extreme example of the action of IgE is acute **anaphylactic shock**. Certain individuals can become exceptionally sensitive if exposed to some antigens. Eventually they may suffer an acute life-threatening reaction when they meeet the same antigen again. Such hypersensitive people may react to such things as penicillin, an injection of an immunoglobulin prepared in an animal, or a bee-sting. The sudden respiratory and circulatory collapse can be fatal.

2.4 ARTIFICIAL ACQUIRED IMMUNITY – IMMUNIZATION

In 1796 Edward Jenner inoculated James Phipps with pus taken from a cowpox sore on the hand of a dairymaid, to protect him from smallpox (Box 2.3). This laid the scientific basis for the production of artificial acquired immunity, to be distinguished from the natural variety (Figure 2.1). The technique was called **vaccination**, from the Latin word for a cow. When Pasteur began to immunize animals and people by artificial means he called the process vaccination, in honour of Jenner. The antigens used for vaccination are known as **vaccines**. Today many different vaccines are available to **vaccinate** (immunize) people against a number of infections. In each case an antigen derived from the microbe responsible for the infection must be introduced into the body. This may be by mouth, by inhalation or through the skin by scratching or, more commonly, by injection.

Box 2.3 Vaccination

Jenner's paper on vaccination was turned down by the Royal Society for 'lack of proof'. He published it at his own expense. It was read by Benjamin Waterhouse, Professor of Physic at the new medical school at Harvard, USA. He vaccinated his own son, and six of his servants. President John Adams refused to support the more widespread use of the vaccine, but the presidential candidate, Thomas Jefferson, was in favour of it. In 1802 in Boston 19 boys were vaccinated. Later they were innoculated with the smallpox virus, and were found to be immune. Two unvaccinated boys given the same virus developed smallpox. The practice of vaccination spread rapidly in the USA. In Europe even Napoleon had his soldiers treated with 'le vaccin jennerien'. The first child to be vaccinated in Russia was named Vaccinof!

The antigenic material used in a vaccine may be separated from the microbe by some chemical or physical process. In some cases it is a toxin that is the pathogenic trade mark of the microbe. If so the toxin is first made into a non-toxic but still antigenic **toxoid**. Vaccination with a toxoid produces antibodies called **antitoxins**. Alternatively the antigens may be still attached to the bodies of fully pathogenic microbes. These have been killed, of course, to avoid transmitting the infection, but this is done with care to make sure first, that they are certainly dead, and second, that the vital antigens are still active. This product is a **killed vaccine**. In some cases live microbes are used (**living vaccines**), though they have been altered (attenuated) in the laboratory to make them less pathogenic and so reduce or eliminate their capacity to cause disease. Immunization with a live vaccine results in a true infection, but one that is trivial or asymptomatic compared with the natural one. This was Jenner's approach, though in his case the virus did not need to be attenuated. The cowpox and smallpox viruses are very similar and they share many antigens. In humans, the cowpox virus produces a small local reaction compared with the severe general and potentially lethal infection that, at one time, followed an invasion by the smallpox virus. This first vaccine was so successful that, after 181 years, smallpox was the first infectious disease to be eradicated.

Vaccination stimulates the immune system of the person immunized to do what is necessary so the result is **active immunity**, although produced artificially. Like its natural counterpart the immunity is more or less long-lasting, but 'booster' doses of vaccine may be needed to establish initial protection at an adequate level, and to maintain it. Only a limited amount of antigen can be introduced at one time with toxoids and killed vaccines, so repeated injections are usually necessary. With a living vaccine the microbe multiplies in the tissues of the person vaccinated. Quantities of antigen are produced on-site, so repeated doses may not be required.

Passive immunity may be induced by the injection of antibody made in the laboratory, or that has been extracted from the blood of another individual. This may be a human or an animal in whom the type of antibody required was formed artifically by vaccination, by a natural infection, or by vaccination to boost pre-existing naturally acquired immunity. Care is taken to preserve the delicate protein antibody (immunoglobulin or **immune globulin**) and at the same time to exclude infective agents that might be present in the serum of the donor. Precautions against the latter risk have not always worked, and serious infections have followed the use of immune globulins that contain microbes, particularly viruses. Hepatitis B has been transmitted in this way, indeed the old name for it, 'serum hepatitis', acknowledges this. The human immunodeficiency virus (HIV, the virus responsible for AIDS) is a new threat. Passive immunity produced by the injection of pre-formed antibody is short-lived (perhaps six months) because the factory for the production of more of it is elsewhere. The only way recipients can make their own antibody is by exposure to the relevant infection, or after vaccination. When an individual protected by antibody is infected by the equivalent natural disease the infection that develops is likely to be mild or subclinical, but the result is now an active immunity. This is what happens in babies for so long as they are protected by antibody derived from their mothers.

2.5 CONCLUSION

A great deal more is known about the mechanisms of immunity than appears in this outline. In particular the physics and chemistry of many of the reactions described here in a general way have been worked out in some detail, though much remains to be discovered. The subject is developing rapidly in many parts of the world and confusion can arise when new components of immunity are discovered simultaneously in several places, where they may be given different names and allocated different functions. In the end they may prove to be a mixture of several factors, each with a different role. For the ordinary onlooker, however, there are plenty of sources of reliable and up-to-date information.

FURTHER READING

See Appendix B.

Microbial aggression | 3

3.1 INTRODUCTION

A few highly pathogenic microbes have learned to overcome all the defences described in the last chapter. Rather more have discovered how to survive on body surfaces, and these with others may also become pathogens if the defences are weakened in some way. Microbes use a variety of aggressive techniques to achieve pathogenic status. They may resist the defence mechanisms, avoid them altogether, or use counter measures to neutralize or destroy them. They can then live on the surfaces of human bodies as colonists or commensals, or invade their tissues as pathogens. Words like 'attack', 'invade' and 'defend' used to describe these relationships suggest a conscious, warlike process. They are in fact extensions of the rule often quoted as 'nature abhors a vacuum'. Wherever there is space to be exploited, some living thing will evolve to occupy it, and sooner or later the site becomes a 'battleground' as rivals compete for it. The human body is just another such site.

Many plants and animals produce chemical poisons or toxins. They are used for defence (the sting of a nettle, for example) or to kill or immobilize prey (the venom delivered by the bite of a snake or spider). Microbes also elaborate toxins and these play an important part in their relationships with each other and with their larger hosts (Box 3.1).

Box 3.1 Poisoned arrows

The word 'toxin' is derived from the name of the poison Greeks used on their arrows. It was applied to bacterial products by Pasteur's greatest pupil, Emile Roux, and his assistant, Alexandre Yersin. They discovered that broth in which the diphtheria bacillus had grown was still poisonous to animals after it had been filtered and freed of all the bacteria. They called the poison a toxin.

3.2 RESISTING THE DEFENCES

Acid is one of the main antimicrobial defences deployed at the surface of the body. The level of acidity in the stomach is sufficient to kill most microbes, but elsewhere it is less strong. Some microbes survive in mildly acid surroundings that inhibit others. Staphylococci can multiply on the skin in a fairly acid and relatively dry environment that other bacteria find inhospitable. They have even learnt to use the fatty acids secreted onto the skin as a source of food.

3.3 AVOIDING THE DEFENCES

A microbe avoids the defences altogether if it produces a toxin outside the human body. This strategy is successful if the toxin can find a route by which to enter an individual to cause disease. Some of these toxins are among the most powerful poisons known, with effects that range from the mildly unpleasant to the devastating. Some of them reach us through food. We cannot avoid sharing what we eat and drink with microbes, bacteria and fungi in particular. Food 'goes off' or becomes 'bad' when microbes eat it before we do, though in some cases (for example alcohol, vinegar, cheese, yoghurt, some sausages and soya-bean and fish sauces) the process of 'going off' is employed to make the product more useful or attractive. A microbe that multiplies in food may produce a toxin and when enough has been formed the food is poisonous. If it has not become unpalatable at the same time and it is eaten, the result is **food poisoning** (Box 3.2 and Chapter 28).

Box 3.2 Poisoned food

Some years ago salmon canned on an island in the North Pacific was accidentally contaminated with *Clostridium botulinum*. These bacteria multiplied inside the cans and as they did so they produced botulinum toxin. The cans eventually reached England and some people who ate their contents developed botulism and died. Other bacteria, for example *Staphylococcus aureus*, *Clostridium perfringens* and *Bacillus cereus*, may do the same thing when they multiply in food, though the effects of their toxins are generally less severe. They cause diarrhoea, vomiting and abdominal pain. Any mixture of these symptoms is called food poisoning (or gastroenteritis or enteritis according to which symptoms predominate).

Collectively these poisons are called **exotoxins** because the bacteria concerned secrete them actively into their surroundings. In the example given the surroundings are the food in which they multiply. Some of these toxins are destroyed by heat, so if food is well heated or re-heated and is eaten while still hot, harm may be avoided. Exotoxins that act in the gastrointestinal tract to produce gastroenteritis are also called **enterotoxins** and those that attack nervous tissue, **neurotoxins**. Other exotoxins damage other organs.

The exotoxins mentioned so far are produced at some distance from their victims. The separation may be measured in metres or in thousands of kilometres, so the bacteria concerned completely evade the mechanisms of human immunity. All that is required is for the pre-formed toxin to be in an active state when it reaches a victim, and that there exists a route by which it can enter his or her body.

Some exotoxins operate over much shorter distances. The microbes concerned come within range of host defences before they can deliver them, to cause disease. Cholera and diphtheria are examples of the short-range actions of exotoxins. The bacteria responsible remain on the surface of the body, so only need to overcome the superficial defences described in Chapter 2. To do this they must survive and multiply in the conditions that exist at the site of their initial attack.

Vibrio cholerae must survive the acid secreted in the stomach to reach its target in the intestine. When the vibrios manage this (Chapter 11) they adhere firmly to the cells that line the intestine where they multiply and produce an enterotoxin. This enters the cells to which the bacteria are attached and disrupts their activities. Depending on how many cells are attacked the result varies from a mild to a copious and potentially lethal diarrhoea. The vibrios must adhere firmly to the cells otherwise they would be washed away in the first flood they cause, before much harm is done. *Corynebacterium diphtheriae* has easier access to its usual target in the throat, but once there it operates similarly. The toxin is absorbed through the mucous membrane on which the bacteria multiply and enters the circulation to attack the heart and nerves.

The toxin of *Clostridium tetani* also operates at short range, but the bacterium adopts a different strategy. Its spores enter the body when an accident causes a dirty wound. Some of the surrounding tissue may be killed at the same time, or it may die later. Immune mechanisms do not operate in dead tissue so when the tetanus spores germinate within it the bacteria that emerge multiply without restraint and produce their neurotoxin. This diffuses into the surrounding living tissues to reach the circulation to attack the nervous system and cause tetanus.

Box 3.3 Exotoxins

All exotoxins are proteins, made up of two parts. Part 'B' binds to the surfaces of target cells so that the toxic part, 'A', can enter them to cause damage. Part 'B 'determines which kinds of cells are attacked, and so what symptoms are produced, but by itself it is usually non-toxic.

Many microbes have been found to produce exotoxins, and more undoubtedly await discovery (Box 3.3). All of them can act at a distance from the bacteria that produce them, even when these multiply inside rather than outside the body. If the bacterium *Streptococcus pyogenes* penetrates into the tissues of the throat it multiplies there to cause tonsillitis. At the same time it may produce an **erythrogenic toxin**. Unless the victim has met the toxin before and is immune to it he or she will then develop the generalized red rash of scarlet fever as well as a sore throat. The exotoxin produced in the throat is the cause of the rash. There are no streptococci in the skin.

The counterpart of an exotoxin is an **endotoxin**. Endotoxins are toxic moleculers that are parts of the structure of a microbe. They express their toxicity locally inside a host if they are exposed on the surface of living attackers, or more generally if they are released from inside them when the microbes die and break up to reveal the toxin. Their activities do not compare with the more dramatic exotoxins but endotoxins have some important properties. Some of them trigger the process that raises the temperature of the body to cause fever and they may stimulate the production of C-reactive protein.

The defences of the body may also be avoided, at least in part, if an invader enters one of the cells of its host. In this way it evades all the humoral defence mechanisms. This is the strategy adopted by viruses, by some bacteria (in tuberculosis, leprosy and listeriosis, for example) and by some protozoa. Most cells are poorly protected against an attack that develops from within them, but the situation may be remedied when the invaded cells are destroyed together with their foreign contents by NK or activated 'T' cells (Chapter 2). Phagocytic cells are specifically designed to ingest and destroy microbes, but even they may be subverted. Some microbes contrive to stay alive after they have been engulfed by a phagocyte and enclosed in its vacuole. Some have learnt to prevent the discharge of the toxic mixture designed to kill them, or to resist its action, or to avoid it altogether when they escape from the vacuole into the cytoplasm of the cell. Not only can these microbes multiply inside cells that should kill them but they are carried by the cells to distant parts of the body and so spread the infection.

3.4 COUNTERING THE DEFENCES

If the defence mechanisms of the body are not resisted or avoided they may be countered. Many pathogenic microbes are equipped with extra overcoats or **capsules** (Chapter 5), and all of them are covered by a cell wall or a limiting membrane of some kind. Some microbes make these coats from molecules that resemble those normally found in their hosts. Invaders not recognized as foreign escape attack because the innate defences are not activated. The adaptive immune system is less easily fooled but as it takes time to develop a response the invader is given some leeway.

Some microbes even counter the action of the adaptive system. To accomplish this they change the chemical structure of their outer coats from time to time. When the host's 'B' cells have recognized an invader as foreign and have wound themselves up to do something about it, the effort is wasted. The wily microbe, confronted by an effective defence, promptly adopts a new disguise. This happens in relapsing fever (Chapter 14) and in trypanosomiasis (Chapter 25) in which the deception is repeated a number of times in the course of a single infection. The same thing can also happen outside the body, between successive infections. Humans may suffer from more than one attack of influenza A. This is because the outer membrane of the virus is changed periodically. Although essentially the same microbe, its new outer coat is not recognized by defences that would otherwise have excluded it.

An infection may be snuffed out in its early stages by the innate immune system. Activated phagocytes are important in this process. Bacterial capsules or the other surfaces of pathogenic microbes may be provided with mechanisms that slow down or stop this activation. Although an initial inhibition of phagocytosis is usually corrected by the adaptive immune system, a good deal of damage may be done before this can happen. The activation of phagocytes by the adaptive system may also be countered. Some microbes (staphylococci and streptococci for example) are equipped with surface materials that can block or divert the defensive activities of humoral antibodies (Chapter 6).

Microbes have other aggressive qualities. Even when a host uses the blood clotting mechanism to prevent the spread of an infection this defence can be neutralized by microbes that produce the enzyme fibrinolysin, which dissolves the clot. Other microbial enzymes dissolve the natural 'glues' that hold tissues together. Microbes that produce these enzymes cause infections that spread more easily. Another toxic enzyme (**leucocidin**) can attack and destroy the phagocytes before they can come to grips with the microbes that produce it.

This array of aggressive mechanisms is opposed by the human immune systems described in Chapter 2. The narrow margin that usually favours the human race depends on a fully competent immune system that works at full efficiency. Relatively small defects lower the defences and open the way to infections. Those who work in professions concerned with human health need to be aware of this. Their patients' immune systems are to be preserved intact, so far as this is possible. Unfortunately damage is a common and often inevitable consequence of treatment. It is necessary to strike a balance between the good that can be done and the harm that may result, and when the choice is made in favour of treatment to take whatever precautions are available to limit the damage.

3.5 LOSS OF AGGRESSION

A microbe may be alive or dead. When alive some of them can be harmful but when dead they may be thought harmless. This is an error because the existence of endotoxins means that dead microbes, in sufficient numbers, are still poisonous. Medical products that are introduced into the body must be sterile to avoid the entry of living microbes but sterilization does not destroy any endotoxins that may be present. In this context endotoxin is the **pyrogen** that is responsible for an unpleasant feverish reaction when something that contains a small quantity of it is injected. If introduced in larger amounts it can cause severe, perhaps lethal, shock. The manufacturers of intravenous solutions and of the equipment used for transfusions and infusions must pay particular attention to this because there have been tragic accidents when proper care has not been taken.

In normal circumstances animals and plants begin to deteriorate and die over a significant period of time. Vital functions are lost one by one. The same is true of microbes, though the effect of this on their aggressive function is widely ignored. When microbes are deliberately grown in the laboratory they are offered every facility to multiply. Anything that might frustrate this is carefully excluded, including of course the mechanisms of immunity. This is not the situation microbes meet in the real world. Other than in special circumstances they have to operate in the presence of active immune mechanisms to express their pathogenicity. It is therefore possible to recover pathogens in the laboratory whose aggressive qualities are diminished or absent. Despite this when an apparently pathogenic microbe grows in the laboratory so shows that it is alive it is widely assumed that it is also in possession of its full pathogenic potential. Some workers have explored this twilight zone between microbial life and death, but their work has not attracted much attention (Rammelkamp et al., 1964; Hinton et al., 1960; Holloway et al., 1986; Rotter et al., 1988).

Streptococci taken from the throat of a person with tonsillitis reproduce the disease in human volunteers when they are transferred directly from one throat to another. If the same bacteria are exposed to the air in the absence of water they quickly lose this ability, though they can still grow readily on laboratory culture media. They are alive, but have lost their virulence (Box 3.4).

Box 3.4 Contaminated blankets

Streptococcal sore throats can be a serious problem in camps where military recruits are trained. Experiments were done to find out how the streptococci spread, and how to prevent it. In an early investigation unlaundered blankets used by recruits just before they were admitted to hospital with sore throats were reissued to 85 new arrivals. They were heavily contaminated with streptococci. Another 177 men in the same barracks were given freshly-laundered blankets, that were free of streptococci. In the days that followed there was no difference in the incidence of sore throats in the two groups. Although the streptococci on the dirty blankets were still alive, they had lost their ability to cause infections. Other experiments confirmed this conclusion and showed that partly dried bacteria are no longer virulent.

The same thing happens with staphylococci, as was shown by an experiment using animals rather than human volunteers. Organisms damaged by disinfectants also lose some of their virulence before they finally die. Many pathogenic microbes depend for their aggression on mechanisms sited on their surfaces. At this superficial level damage caused by desiccation or weak disinfectants will be noticed some time before vital internal structures are involved and the microbe finally dies.

Failure to give due weight to this factor is evident in the prominence given to routes of infection that involve microbes in the air or on surfaces where they are exposed to the effects of dessication. This is discussed further in the next chapter.

FURTHER READING

1. Rammelkamp, C.H., Mortimer, E.A. and Wolinsky, E. (1964) Transmission of streptococcal and staphylococcal infections. *Annals of Internal Medicine*, **60**, 753–8.

2. Hinton, N.A., Maltman, J.R. and Orr, J.H. (1960) The effect of desiccation on the ability of Staphylococcus pyogenes to produce disease in mice. *American Journal of Hygiene*, **72**, 343–50.
3. Holloway, P.M., Bucknall, R.A. and Denton, G.W. (1986) The effects of sub-lethal concentrations of chlorhexidine on bacterial pathogenicity. *Journal of Hopsital Infection*, **8**, 39–46.
4. Rotter, M.L., Hirschl, A.M. and Koller, W. (1988) Effect of chlorhexidine-containing disinfectant, non-medicated soap or isopropanol and the influence of neutralizer on bacterial pathogenicity. *Journal of Hospital Infection*, **11**, 220–5.
5. See Appendix B.

4 Infection

4.1 INTRODUCTION

An infection exists when a microbe strikes up a relationship with a host and there is a reaction between them. The reaction may be the appearance of disease (a clinical infection) or it may be less obvious and detected only by a laboratory test (a subclinical infection). When a microbe and a host live together with no reaction between them, the relationship is a colonization. It has been pointed out that very large populations of microbes live on the human colonized surfaces, as commensals. These are **normal colonizations**, and a healthy commensal flora provides its host with an additional defence against attack by other microbes.

An **abnormal colonization** is the result of some disturbance of the normal state. When the normal flora is upset one component of the former healthy mixture may become dominant, or the upset may follow the arrival and establishment of a new, atypical colonist. This newcomer may be a member of the normal flora at another surface of the same body, for example a microbe that normally lives in the intestine may establish itself on the skin. Alternatively the newcomer may be a complete stranger, both to the individual and the site. An abnormal colonization may develop as the result of injury, disease, the use of drugs or anything else that alters the physical or chemical state of the surface. It may involve one part of the body, or the whole of it.

If the defences that preserve the privileged surfaces fail or are bypassed, an abnormal colonization is established in what was, and should be, a microbe-free zone (Chapter 2). The situation is now abnormal even if the microbes concerned are 'harmless' members of the host's normal flora. An object that pierces, cuts or tears the surface of the body may carry microbes directly into the internal tissues. Even when this invasion does not happen at once the wound opens a path along which members of a local normal or abnormal flora can travel to reach the tissues some time later. Even after microbes have reached the tissues their presence remains a (very abnormal) colonization until the body reacts to their presence.

New colonists may not be as friendly as the old ones. The regular use of personal deodorants that contain antiseptics may cause discomfort because of this. An antimicrobial drug often disrupts the normal flora of a patient at the same time as it attacks the microbe for which it has been prescribed. For example ampicillin may disturb the flora of the gut and diarrhoea is a common consequence of its use, particularly when the drug is taken by mouth.

For these and other reasons a patient in hospital often develops a new abnormal flora in the days after admission. Some time later these new residents may cause a hospital-acquired infection.

These are not the only ways in which treatment can be harmful. A surgical operation produces a wound, with the potential for infection already described. A urinary catheter bypasses the defences of the urethra. An endotracheal tube blocks the top of the mucous escalator, and anaesthetics and some pain-killing drugs paralyse the cilia that keep it moving. Some unwanted results must be accepted as the price of doing good, but a proportion of the harm can be avoided if appropriate precautions are taken. These matters are the concern of the sub-speciality that deals with the control of infection in hospitals.

An abnormal colonization may be corrected and the normal healthy state re-established without an infection, or the correction may involve the activation of the immune system. The rapid response of the innate system may be all that is required, or if not, the more powerful adaptive system is induced to produce antibodies and activated 'T' cells (Chapter 2 and Box 4.1). The

---

Box 4.1 Antibodies

Initial developments in microbiology and immunology were focused on two great men in two great European cities. Louis Pasteur in Paris and Robert Koch in Berlin each attracted a band of disciples who came from far and wide to work with the masters. Two of these men made contributions to the early development of immunology. Elie Metchnikoff, slighted by Koch, settled in Paris with his theory of cellular immunity (Box 2.1) while Emil Behring (1854–1917) made discoveries in the humoral field, in Berlin.

Behring, like Koch, was one of 13 children and also had a military background. In 1889 he joined Koch in Berlin, where the diphtheria bacillus had been discovered and just a year after diphtheria toxin had been identified in Paris. Behring found that minute doses of the toxin injected into guinea pigs protected them from a lethal dose given some time later. Not only that, but serum taken from an immune guinea pig protected a previously un-immunized animal from a lethal dose, and might even save the life of one already suffering from diphtheria. This demonstrated the existence of antibodies and of humoral immunity, and led to the treatment of diphtheria (and tetanus) with antitoxic sera made in animals and eventually to diphtheria and tetanus vaccines.

---

infection is subclinical until microbial invasion causes disease. An abnormal colonization always precedes the appearance of a clinical infection. The duration of this abnormal colonization is the **incubation period** for that infection. Incubation periods can last for hours or many years, though with most infections they are measured in days. In the latter part of an incubation period, the colonization becomes a subclinical infection before symptoms begin to appear (if they do) to convert the subclinical state into a clinical one.

To express aggression against a human host a microbe must pass through several hoops before the stage of an abnormal colonization or an infection is reached. First the microbe must have a **source**. This must be connected to the human victim by a **route of transmission** along which the microbe travels. The route must lead to a **point** (or **portal**) **of entry** on or in the victim.

The microbe must then multiply and overcome the victim's defences (this is the incubation period) and finally cause an infection (Figure 4.1). There

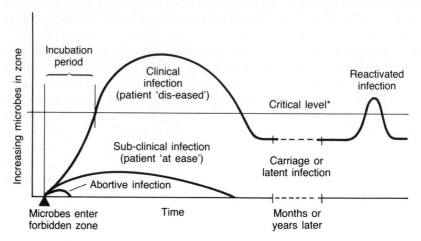

Figure 4.1 The course and outcome of a microbial invasion of the 'forbidden zone' (Figure 2.1) to illustrate the features of clinical, subclinical and reactivated infections, and carriage. The 'critical level' marks the boundary between subclinical and clinical infections.

are two other factors critical to the outcome. First, microbes must reach their intended victim in numbers greater than the **minimum infectious dose (MID)**, and second their aggressive qualities must be unimpaired. Although this series of events applies to all infections there are major differences in emphasis and detail. For example, the early stages are abbreviated when a normal microbial inhabitant of a body surface invades its host to cause an infection. A pre-formed exotoxin has completed its stage of production (as a result of microbial multiplication) before the toxin reaches the victim's tissues.

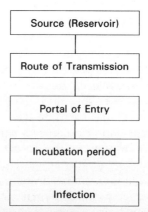

Figure 4.2 The much-quoted 'chain of infection' that takes no account of variations in the minimum infectious dose of a microbe, or the loss of its virulence as it traverses the chain. It also ignores the major factor of host resistance.

The events that commence with a microbe in a source and end with a human infection are often depicted as a chain, with each event represented by a separate link (Figure 4.2). The attraction of this simile is the idea that a 'weak link' may be detected. If this is broken the chain fails and infection is prevented. This simile does not readily accommodate the requirement for a minimum infectious dose or the need to preserve microbial aggression. The outcome is also heavily influenced by another variable, the competence of a potential victim's immune system. These added complexities explain why, in practice, weak links have proved elusive.

One other point must be made in this introduction. Infection has been defined as an interaction between two living things. An inanimate object does not react to the presence of a microbe so cannot be infected. Yet such things as infected linen, dressings, or waste are often mentioned. These things may be contaminated by microbes, but they are not infected. A similar error is found in the use of the word **sepsis**. Sepsis equates with infection, with over-tones that suggest the production of pus. The word should only be applied to living things.

Asepsis should mean the absence of sepsis. As often used today it implies the absence of microbes, as in aseptic surgery or an aseptic procedure. The absence of microbes is incompatible with the presence of humans. It may be too late to correct this seeming misuse, but it is useful to be aware of it. A lot of money is spent handling the 'clinical waste' that emerges from hospitals and health centres. Part of the reason for this is that clinical waste is widely perceived as 'infected', so must be handled with elaborate care. This is wrong, and many of the precautions taken as a result are an unnecessary waste of money (Chapters 31 and 32).

4.2 SOURCES

It might seem superfluous to say that to cause an infection a microbe must have come from somewhere, but the idea becomes more complex as it is examined. For convenience, the sources of infections are divided into three categories. Microbes that cause infections may have originated in the same individual, or come from someone else, or from something in the environment. This divides infections into **self-infections**, **cross-infections**, and **environmental infections**, respectively.

The first requirement of a source of infection is that it contains the microbe in a fully virulent state, and in sufficient numbers. At least one MID must be available in the source, and be transportable along the route that leads to the victim. The example of cholera is used to illustrate this. To cause disease several million cholera vibrios must be swallowed if a few are to escape the acid usually present in an active healthy stomach so as to reach the intestine alive. The number present in the largest drinkable volume of even the most heavily contaminated water is quite insufficient. If everybody had plenty of acid in their stomachs all the time, water-borne epidemics of cholera would be impossible. Cholera vibrios can multiply in food but not in water. After a few hours a small number of vibrios in an item of food are converted into a much larger number, eventually in excess of the MID required to over-come the most acid stomach. The source of infection in cholera may be water when stomach acid is deficient, or food when it is not. In an epidemic both sources may be involved. The food may have been contaminated by vibrios that originated in water, but if food causes a case of cholera then this is the primary source of the infection. If the vibrios in the food originally came from water, then the water may be called a secondary source.

Environmental sites are often contaminated by the microbial fall-out from people with infections or colonizations in the neighbourhood. If these microbes play no further part in the onward transmission of infection, they form a reservoir of the organisms concerned. Wasteful and inappropriate activity will follow if proper distinction is not made between sources and reservoirs. Water in the waste traps of handbasins contains microbes washed from the hands of individuals who have used them. These microbes were formerly part of the resident flora of the skin plus any short-term passengers the hands have picked up. In a hospital ward the latter will include bacteria that are the causes of infections among the patients under treatment. At one time some people assumed that the bacteria in the waste traps could escape to cause infections. How they did this in sufficient numbers and by what route they reached patients was not demonstrated. Despite this small electric heaters were fitted to the waste traps of sinks in clinical areas of some hospitals, and the water in the traps was boiled from time to time. This astonishing waste of money arose because those concerned did not distinguish between the real sources of infection (usually direct transfer between patients on unwashed hands) and a harmless reservoir of bacteria (the sink traps).

Another example of this error concerns the bacterium, *Pseudomonas aeruginosa*. This may cause serious infections, but it can also grow in the water in a vase of flowers. Some people were impressed by this and felt that flowers should be kept out of wards. Common sense prevailed when it was realized that there was no route that connected this 'source' with a potential victim, short of using the water as a mouthwash or pouring it into a wound. This 'source' was another harmless reservoir.

The second requirement for a source of infection is that it must allow microbes to retain their aggressive qualities. It was pointed out in Chapter 3 that microbes may lose their virulence some time before they die. Microbes stranded without water find themselves in difficulties. They cannot multiply, and as they lose their own water to the atmosphere they suffer progressive damage until they die. Microbes vary in the speed with which they lose water. For most of them the time taken to die is measured in minutes or hours, unless they are surrounded by materials that protect them from drying out. If they are em-bedded in protein-containing body fluids microbes are preserved just as plants are kept alive in dry conditions by water held in the soil by peat or compost.

Some microbes have evolved methods that slow down the loss of water when they find themselves in dry surroundings. They may be provided with special semi-waterproof outer coats. This approach has been adopted by the pox viruses, and the bacteria that cause Q fever and tuberculosis. An even more effective way is to abandon the vulnerable microbial body altogether, and for the vital parts to retreat into a structure called a **spore**. This is equipped with special coats that are virtually impermeable to water. As the seeds of annual plants survive a winter that kills their vegetative parents, so spores survive and then germinate to recreate the original 'vegetative' forms when suitable conditions return.

Pathogenic microbes are at their most aggressive when they are multiplying in an infection or when they grow actively on an appropriate body surface. Microbes shed into the environment from these sites nearly always find themselves in less favourable surroundings, separated from their accustomed source of nourishment and perhaps deprived of water as well. In most cases they have started the process that ends in their death. At one time hospitals used large quantities of disinfectants to destroy these largely moribund microbes that microbiologists or others had discovered lurking in the dry environment, on walls, locker-tops, floors and so on. In most places this wasteful practice has been stopped and there have been no epidemics of infection as a result. The message is clear: microbes in the dry environment, even when they are recognized as pathogens in a laboratory, have a limited capacity for doing harm. Dry dust that has lain for some time contains very few living microbes, and a question-mark hangs over the pathogenicity of any survivors that are not specially equipped to withstand desiccation. It is important to note that wet environments are quite a different matter.

The microbes that can come from environmental sources to cause infections fall into three classes. As indicated, some pathogenic microbes form spores or are otherwise equipped to withstand desiccation. A few pathogens are normal inhabitants of wet places and some microbes shed from the body can survive for a time or even multiply in a wet inanimate environment. This can be important in hospitals where diagnostic or therapeutic equipment is frequently wet, by accident or design. These pieces of apparatus can provide liquid highways along which microbes travel to enter the bodies of patients. Many microbes that would be harmless if swallowed or painted onto the skin are lethal if they are infused directly into the circulation.

4.3 ROUTES OF TRANSMISSION

There are many routes along which microbes may travel between their sources and their victims. These vary in length and complexity. A baby girl's urinary tract is sometimes invaded by the *Escherichia coli* that abound in her faeces. The defences mounted by her short urethra may be overcome by the weight of the challenge. The microbe has almost no distance to travel to reach her bladder, so the route is short. Such short routes are characteristic of most self-infections, where the source of the infecting microbes and their victim are the same person. The largest number of microbes found closest to an individual are the members of that individual's normal (or abnormal) flora. This is why self-infections are so common. At the other extreme for distance is the case of botulism mentioned earlier, when the source itself (cans of salmon) travelled a route measured in thousands of kilometres to reach their victims. Routes vary in complexity. At one pole is the simple movement of the microbe described in the baby girl. At the other are the extremely tortuous routes followed by microbes that multiply as they pass through insects, snails or other animals before they reach and cause infections in human hosts.

4.3.1 Routes in self-infections

In the case of self-infections the route taken is often by **direct extension**. In this case microbes travel along the route with no more assistance than might be provided by the normal movements of their host. These movements may release microbes from their usual home and transfer them onto a privileged surface (such as the urinary tract or the lung) or through a wound into the tissues. Alternatively the route may be by **indirect extension**, when the transfer is made by other agencies. These may be hands or inanimate objects that belong to or are used by the same individual, or by a third party. This provides a common route of infection in hospitals and other health establishments. Doctors, nurses and others may touch, or apply instruments to, more than one part of the anatomy of a single patient. Although hands and inanimate objects are more or less inhospitable, the microbes they pick up do not suffer damage or die in the few seconds taken to transfer them from one body site to another.

Patients in hospitals may be the victims of self-infections with microbes that are recognized as hospital residents. In many cases these have been acquired after admission to hospital and have become part of an abnormal colonization before something happens that allows them to cause an infection (Box 4.2).

Box 4.2 New colonists

In the community about 20% of people carry small numbers of *Pseudomonas aeruginosa* in their gastrointestinal canals. After a few days in hospital the proportion may rise to 30–40%, or even to 90% after seven days in an intensive care unit. This is why *Ps. aeruginosa* causes more infections in hospitals than in the community.

4.3.2 Routes in cross-infection

The term **cross-infection** is applied to an infection with a microbe that has recently come from someone else. The boundary between this and a self-infection requires careful definition. This is because a clinical infection is always preceded by an incubation period, that is by a colonization and a subclinical infection with the same microbe. If a colonization has existed for longer than the incubation period, a clinical infection that follows is counted as a self-infection derived from a pre-existing abnormal flora. If not it is a cross-infection. Difficulty arises because in any given case it may be difficult or impossible to establish the exact length of the incubation period.

The most common route for the transfer of microbes between people is by contact. This may be direct, as between members of families, lovers, friends, or in a crowd. Hospitals and other medical establishments are places where very intimate contacts are made between perfect strangers. From the point of view of a microbe the transfer is unsuccessful if it fails to establish itself on or in a new host. This is the most common outcome. If the microbe succeeds it may become a new member of a normal flora, or it may establish an unhealthy abnormal colonization, the latter perhaps a prelude to an infection. Transfer by direct contact is the most effective of the routes associated with cross-infections. This is because it allows the movement of microbes between individuals in large numbers, without an intervening period in which they might suffer damage. For this reason health care workers with infected hands or who for some other reason are heavy carriers of pathogenic microbes are a major hazard to patients.

Microbes may also be transferred by indirect contacts that involve third parties. Hands and inanimate objects are the most important vehicles, just as in the case of self-infections. Health care workers often move rapidly from patient to patient with hands that touch or instruments that are applied. Once more the rapidity of movement between patients ensures an efficient transfer. Of course the hazard disappears if hands are washed properly between patients or if the objects used are disposable, or are treated to remove any microbes they have picked up.

In the case of inanimate objects the length of time that elapses between uses has additional significance. Not only does the hazard diminish as the microbes on it or in it are damaged and die when they are deprived of water and nutrition, but there is also a semantic point. What is a cross-infection when the period between uses is short becomes an environmental infection as the period lengthens, and the source of the contamination has been forgotten or goes undetected. A similar distinction might apply to the case of an infection transmitted when one person sneezes in the face of another (an obvious cross-infection) compared with someone who enters a room in which a former occupant sneezed five minutes ago (apparently an environmental infection). The subject of the air as a source or route of infection deserves special attention. It is dealt with in the next section.

4.3.3 Routes in environmental infections

Environmental infections are caused by microbes from sources other than people. This simple statement hides a deal of complexity, and some confusion. Microbes found in the environment fall into one of two broad categories. By far the greater number spend their whole existence in what may be thought of as our inanimate surroundings, but which in fact teem with life. The soil in fields and gardens contains myriads of microbes. Many of them have developed mechanisms that allow them to survive a drought, but none of them can multiply in the absence of water and nutrients. Most are not even at their best when the temperature rises to that of a warm-blooded animal.

Only a few of the many microbes that exist can operate successfully both in the environment and on or in the human body. *Ps. aeruginosa* is one of these. Although the water in the flower vase mentioned above is not a source of infection, a warm swimming pool or spa bath may be. A swimming pool may also be the source of free-living amoebae that occasionally invade the nervous system to cause a severe meningo-encephalitis. Another member of the genus *Pseudomonas*, *Ps. pseudomallei*, lives quite normally in the environment of some parts of the humid tropics. This bacterium is the cause of a serious human infection, melioidosis (Chapter 12).

In most cases environmental microbes reach humans by short and direct routes, when people come into physical contact with their sources. This is what happens when individuals swim or relax in contaminated water. Another obvious route is through food, as in food poisoning. In hospitals, food has an extra dimension as a route of infection. *Ps. aeruginosa* may set up house in the workings of a food blender if these have not been cleaned meticulously. If a contaminated blender is used to prepare a meal for a patient who must be fed through a naso-gastric tube, bacteria that can multiply in the liquefied food are added to it. The patient is then exposed to a large dose of bacteria that are particularly dangerous for someone who is sufficiently ill to need feeding in this way. Even medicines are not immune. Eyedrops that are carelessly produced or contaminated in use may contain *Ps. aeruginosa*. If such drops are applied to an injured eye, infection and blindness can be the result.

Confusion arises because an individual with an infection frequently sheds the microbes concerned into the environment where they can be detected by a simple laboratory test. Before the antimicrobial era the haemolytic streptococcus *Streptococcus pyogenes* was greatly feared. This bacterium can attack apparently healthy people and kill them in a few hours. It is the cause of puerperal sepsis, an infection in the wound left when the placenta separates after childbirth. At one time this infection took a heavy, tragic toll among young women. Bacteriologists discovered that the bedding, floor and other surfaces and objects near such patients were plentifully contaminated with streptococci. Practices were designed and introduced to prevent the spread of infection from these apparent environmental sources.

In the 1940s it was discovered that streptococci exposed to the atmosphere rapidly lose their aggressive qualities (Chapter 3). This should have led to a re-examination of these practices, but penicillin had just arrived and the defeat of the streptococcus seemed assured. The practices were not modified as they should have been and a little later the same precautions were adopted to prevent the spread of the microbes that replaced streptococci as the leading causes of infections. Some of these precautions, originally introduced to protect against streptococci, are still used today in an attempt to prevent the spread of quite different microbes. The efficacy of these measures is more a matter of opinion than of scientific proof. This is a pity, as they cost money.

Complexity among environmental infections is a consequence of the existence of other forms of life. An infection that is primarily one of animals which may by accident be transferred to a human host is called a **zoonosis**. The microbes concerned reach a human victim by a variety of routes, either by direct contact with an animal, through animal products such as food, wool, hair, hide or bones, or through water contaminated by it, or through the air, or with the help of a blood-sucking insect. A single microbe may make use of several of these routes at different times and in different circumstances. Further descriptions and examples are given in later parts of this book, but routes involving blood-sucking insects and the air require more consideration here.

4.3.4 Blood-sucking insects

An infection may be carried from one victim to another as a by-product of the feeding habits of bugs, lice, midges, mosquitoes, flies, fleas, ticks and mites. Blood-sucking arthropods may all be referred to conveniently, though somewhat inaccurately, as insects (Chapter 27). The microbes responsible are acquired when an insect draws a meal of blood from a human, or from some other animal. The source of the blood meal must be suffering from an infection with a microbe that is present in its blood in such quantity that the average insect-full contains at least one MID for the insect. The microbe must be able to multiply, this time in a cold-blooded host that may or may not appear to suffer as a result. When sufficient new microbes have developed in the insect they are ready to be transmitted to the source of its next blood meal. They may enter this new victim in the insect's saliva or through insect droppings that are rubbed into the bite or into other abrasions made when the victim scratches.

In these cases the infected humans or animals are reservoirs of the microbe until another insect feeds on one of them, so converting that individual into a new source. The insect host also acts as a source, particularly when they live for a long time, or pass the microbe directly from one generation to another. When the source of a human infection is an animal, the disease is a zoonosis. Most diseases that are spread by insects are found in the tropics. This may be because the insects concerned can only survive at higher temperatures (with dengue and yellow fever, for example), or because the insects or the microbes they carry have disappeared from some cooler parts of the world. This is what has happened with malaria, plague, and the louse-borne forms of typhus and relapsing fever.

4.3.5 Airborne spread

As a major part of our environment the air did not escape the attentions of early microbiologists. In 1864 Louis Pasteur disposed of the theory that life can be generated spontaneously by an experiment that showed that microbes are present in the air. The surgeon Joseph Lister repeated the experiment and concluded that these airborne microbes were responsible for post-operative 'putrefaction' (infection). Lister began to attack microbes in wounds with carbolic acid in 1865, and later used his famous spray to extend the attack into the air. The number of his patients who died after amputation fell from a catastrophic 46% to a still rather disastrous 15% (Box 4.3). Florence Nightingale believed in the importance of fresh air and cleanliness in the fight against infection. Microbiologists examined the air and discovered that it often contained the same microbes as they found in infections and other parts of the human environment. They concluded that the air is another important route for the transmission of these infections. Policies and procedures were designed in an attempt to break this link.

The logic was faulty, for three reasons. First, most of the microbes found in the air belong to species that are poorly or nearly completely

Box 4.3 Lister and antiseptic surgery

Joseph Lister was the second son of a Quaker family. His father, a wine merchant, was also a gifted amateur microscopist who made his own lenses. In 1843 Joseph, who used a microscope extensively in his researches, entered University College, London. He was the first of his family to go to university because, as Dissenters, they were excluded from Oxford and Cambridge. There was nowhere else to go in England until the non-sectarian London University was established in 1828. In 1846 at University College Hospital Lister witnessed the first use of ether as an anaesthetic in England. He became a Fellow of the Royal College of Surgeons in 1852 and in 1853 he moved to Edinburgh as assistant to James Syme, whose elder daughter Agnes he married in 1856. After he had introduced carbolic acid into surgery in 1865 and his spray in 1870 he moved back to London in 1877, to King's College Hospital. Later he changed his mind about the spray and in 1890 when he addressed the International Medical Congress in Berlin he said, 'I am ashamed that I ever recommended it'. At the same Congress, Robert Koch unwisely proposed the use of tuberculin in the treatment of tuberculosis. Lister took a consumptive niece to him, but she died. Lister was created a Baron in 1897, the year of Queen Victoria's jubilee. Lord Lister died in 1912.

non-pathogenic. Second, they are present in small numbers, so any reasonable volume of air does not contain an MID for most infections. Third, as time passes, an increasing number of the microbes present have been damaged by progressive desiccation, so have lost some or all of any pathogenicity they ever had.

Some attempts have been made to measure the influence of the airborne route of infection, and to assess the effectiveness of measures designed to prevent it. When these have included appropriate controls they have indicated that the airborne route is, at most, of small importance. Even the ubiquitous face mask has never been proved to prevent infections. Some experiments have in fact shown they are useless, and it has even been suggested that masks may cause them (Tunevall, 1991; Schweitzer, 1976). It is curious that, in some quarters, so much attention is still paid to the airborne route of infection.

Among infections that might be spread by the airborne route the best candidates are those of the respiratory tract. Colds, influenza, pneumonia and tuberculosis are examples. In these cases the limited number of microbes present in the air might be counterbalanced by the large volumes inhaled. Although it is well known that 'coughs and sneezes spread diseases', experimental support for this common perception is hard to find. Some experiments have in fact shown that infections due to viruses in particular are more easily spread between individuals by physical contact than through the air, even in confined spaces.

Although it is necessary to present arguments that challenge the overrated importance of the airborne route of infection, it would be dangerous to ignore it completely. Some of the experiments noted above have shown that the airborne route is a minor contributor to some infections. Surgery is an example, but here the prolonged emphasis on the air has obscured other more important causes of infection that could have been dealt with more easily and much more cheaply.

The airborne route is of undoubted importance in two sets of circumstances. The first relates to infections caused by microbes that are designed to withstand desiccation. Tuberculosis is perhaps the most important example of this. An individual who suffers from open pulmonary tuberculosis has an abscess cavity in the lung that communicates with the outside world through the respiratory passages. When such a person coughs, soft cheesy material is expelled from the abscess and this is mixed with sticky mucus. Because of their special properties the tubercle bacilli present do not die or lose pathogenicity when they are exposed to the air. They retain their aggressive qualities for some time, either on surfaces or in the air, or as they move between them. This happens when the larger sticky particles that are coughed up dry and break down into smaller ones of a size that can float in currents of air. Any that contain tubercle bacilli may then be inhaled, to start another infection.

Another example concerns some of the fungi. Most of the ones that live freely in the environment are of negligible pathogenicity. They release their seed-like spores directly into the air so they are naturally designed to withstand desiccation. The small but growing number of heavily immunosuppressed people in the population who, like everyone else, inhale spores of the fungus *Aspergillus* can as a result develop life-threatening fungal chest infections.

The second set of circumstances is illustrated by the example of legionellosis. This is a 'new' disease that has emerged as a result of human ingenuity. Air-conditioning systems and modern plumbing are important for comfort and convenience, and are essential parts of modern architecture. The machinery involved can create niches in which *Legionella pneumophila* proliferates to create a reservoir of this bacterium within the immediate human environment. The machinery itself may then convert this reservoir into a source of infection by discharging a mist of water that contains the bacteria. Infections result if these reach the respiratory tracts of immunodeficient people in or near the air-conditioned building.

4.4 POINTS OF ENTRY

When it has reached the end of a route that leads to a new host, the microbe must establish a bridgehead on or in the victim if it is to continue the attack. This bridgehead forms the point or portal of entry. Any surface of the body may be involved and become the site of an abnormal colonization as a prelude to an infection. Alternatively the microbe may avoid the surface barrier altogether if it enters the tissues through a wound. A wound may be accidental or deliberate, and its cause may range from a road accident to the bite of an insect. A significant number result from medical and surgical activity in hospitals, nursing homes, health centres and individual patients' homes.

Certain routes are linked with particular points of entry. The gastrointestinal canal is naturally favoured by food and drink, and the respiratory tract by microbes that travel through the air. Direct contacts tend to set up abnormal colonizations on the skin or mucous membranes. The microbes concerned may then gain access to the tissues through accidental or deliberate wounds, which may be so small that they escape notice.

A point of entry may be set up before an abnormal colonization has developed. Indeed the former may contribute to the establishment of the latter. The entry point of an intravenous catheter is usually damp, so is likely to become the site of an abnormal colonization. The microbes involved can then spread along the track of the cannula to reach the bloodstream. A burn or some other serious injury produces a wound that contains dead tissue. Dead tissues cannot defend themselves from the attack of microbes and they are also an excellent source of nourishment. Microbes can easily establish bridgeheads in fragments of dead tissue from which they can then attack surrounding healthy ones. This is why infections are such important causes of morbidity and mortality in patients with burns or inadequately treated traumatic injuries.

It has already been noted that tetanus may follow if spores of *Clostridium tetani* contaminate wounds that contain dead tissue. In similar circumstances other members of the genus *Clostridium* may cause gas gangrene (Chapter 8). If the right clostridial spores get into a wound in which there is also dead tissue, they germinate and begin to multiply. Powerful toxins are produced that diffuse outwards to bring death to the surrounding living tissues. At the same time microbes attack the dead tissue and produce gas. This inflates and splits open the affected part and the area of gas gangrene spreads with frightening rapidity. If this happens in a limb, urgent amputation may be the only way to halt the process and save life.

A very special portal of entry appears in pregnancy. If a mother-to-be suffers from an infection, the microbe responsible may reach her baby through the placenta. An infection in the infant may kill it outright, or injure whichever groups of cells happen to be multiplying most vigorously at the time. The organs these cells are destined to form may not develop properly so the baby is born with such congenital defects as blindness or mental retardation. This can happen as a result of infections such as rubella (German measles, Chapter 18).

4.5 AGGRESSION, MINIMUM INFECTIOUS DOSE AND DEFENCE

When a microbe attacks a potential victim the outcome depends on the balance struck between these three factors. This is illustrated by the 'scales of infection' depicted in Figure 4.3. The aggressive qualities of the microbe are placed as weights in one pan, the number of them according to the size of the challenge dose. Weights to represent the host defence are put in the other.

On the side of aggression weight is measured in pathogenicity, on a scale that runs from a lightweight non-pathogen to a heavyweight powerful pathogen. Defences vary similarly between complete (heavy) immunocompetence and a progressive loss of weight as immunity wanes. Host immunity is greatly enhanced in an individual who is also immune as a result of natural infection or vaccination with the same microbe. The effect of the MID is

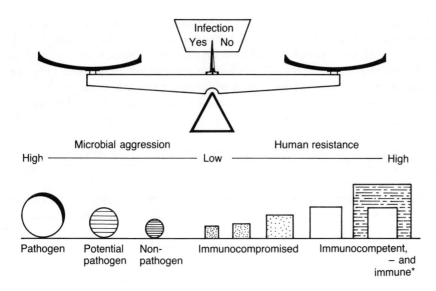

Figure 4.3 The 'scales of infection' on which is decided the outcome of the relationship between microbial aggression and the size of the challenge dose of microbes when these are set against a host's defences. The relative 'weights' of these elements decide between infection or no infection.

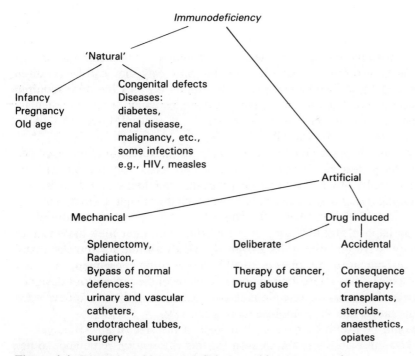

Figure 4.4 Categories of immunodeficiency, with some examples.

measured in numbers of microbes. Smaller numbers favour defence and larger ones, aggression. It has been noted that if three people are exposed to the same risk of infection one might suffer severely, another mildly, and the third, not at all. A little thought will show that each of these three outcomes is determined by varying the way in which the scales are set up. In this case, microbial aggression is the same for all three. Variations in the number of microbes transferred or of the level of a victim's resistance account for the differences.

These considerations are of special importance in the context of health care. The immunity of individuals is reduced when they are sick, or if they are old, pregnant or very young (Figure 4.4). The level of a patient's immunity is nearly always lowered still further by treatment. This has two consequences for the scales of infection. First, as the defences are weakened the weight of aggression needed to overcome them is reduced. The result is that microbes of lesser pathogenicity are made to appear more pathogenic. This is borne out with microbes like *Staphylococcus epidermidis* and *Candida albicans*. Though these are non- or weakly pathogenic for ordinary people they can be killers among patients in oncology and transplant units.

The second consequence concerns microbes of intermediate or high pathogenicity that may cause infections in healthy people. The size of the challenge dose together with any loss of aggressive qualities now determines the outcome in ways that again may be visualized by reference to the scales. The importance of this for patients whose immunity is deficient is that fewer microbes have to be transferred to initiate an infection, that is, the MID is reduced. This means that routes that cannot accommodate an MID of the size required to infect a healthy person may be open to the smaller dose now needed. For example, a healthy person requires a large dose of salmonellas to develop salmonella gastroenteritis. The number is so large that in circumstances of ordinary social hygiene the only way they can be acquired is by eating heavily contaminated food. Among the sick the dose required is reduced. Food becomes toxic when it contains many fewer salmonellas, and hands invisibly soiled with faeces may now transmit infection without the intervention of food at all.

Burns units provide an extreme example of this sort. Burns are occupied by sloughs of dead tissue that lack natural defences. They are quickly and heavily colonized by a variety of microbes that can then spread to infect the surrounding living tissues. Microbes that reach the slough in small numbers are able to multiply in what is the equivalent of a laboratory culture plate. This can happen provided they are still alive even after they have lost all their aggressive qualities. The air now becomes a route for the successful transfer of damaged microbes in small numbers of no consequence in other circumstances. It is interesting to note that many of the precautions designed to prevent the spread of infections in the 1930s and 1940s were developed in burns units, but were applied more generally.

It would be useful to have a clear idea of the size of an MID, yet we know very little about it. One reason for this is a natural reluctance to use humans as experimental subjects. It has been noted that MIDs in cholera and salmonellosis may run into millions of microbes. Direct experiment has shown that at least a million *Staph. aureus* must be injected into the skin to cause a minimal pustule. The injection of fewer than this is without effect and 10 million painted onto intact skin are harmless. By contrast, 100 staphylococci produce an abscess when introduced on a suture. This small foreign body reduces immune competence by a factor of 10 000. It is evident that the side on which the scales of infection come down is determined by quite small changes in the contents of the two pans.

Some work has been done to determine the MID for infection in surgical wounds. Bacteria were counted in each gram of tissue in a wound, or in each millilitre of fluid exudate, at the end of a number of operations. Patients were followed up to see which of them developed infections. The bacterial numbers were compared with postoperative courses. In one experiment the figure of 100 000 bacteria per gram of tissue was judged to separate a low risk of infection from a much higher one. In another experiment the figure was 10 000 (Krizek, 1975; Houang & Ahmet, 1991). It is interesting that these figures are of the same order of magnitude as those that separate colonizations from infections of the urinary tract (Chapter 28).

It must be admitted that knowledge on this subject is deficient. In general, however, it seems that MIDs are large. Alleged sources of infection that contain small numbers of microbes may be viewed with suspicion. The fact that small variations in bacterial aggression and host immunity have such large effect also means that it is necessary to keep an open mind.

FURTHER READING

1. Houang, E.T. and Ahmet, Z. (1991) Intraoperative wound contamination during abdominal hysterectomy. *Journal of Hospital Infection*, **19**, 181–9.
2. Krizek, T.J. (1975) Biology of surgical infection. *Surgical Clinics of North America*, **55**, 1261–7.
3. Schweizer, R.T. (1976) Mask wiggling as a potential cause of wound contamination. *Lancet*, **2**, 1129–30.
4. Tunevall, T.G. (1991) Postoperative wound infections and surgical face masks: a controlled study. *World Journal of Surgery*, **15**, 383–8.
5. See Appendix B.

Bacteria, and the Infections They Cause

PART 3

Introduction to the bacteria 5

5.1 BACTERIAL ANATOMY AND PHYSIOLOGY

Ordinary bacteria range in size from 0.5μm to 10μm in their greatest dimension, so are small even by microscopic standards (Chapter 1). With an ordinary microscope little can be seen of them except their general shape, and (when they do) how they arrange themselves in groups. Two shapes are recognized and, with due allowance for nature's refusal to submit to rigid human classifications, those that approximate to a sphere are **cocci** and the rod-shaped remainder are **bacilli**. The rods may be curved like a comma (the **vibrios**), or appear in the form of tightly or loosely coiled spirals. Sometimes they have knobs on them.

Bacteria reproduce by **binary fission**. At some point as they grow each bacterial cell divides into two. The two '**daughter cells**' may separate at once, or remain together for a time. Cocci that tend to form pairs are called **diplococci**. Some bacteria stay together to form larger groups. If the cells in such a group have divided at random, an irregular cluster of several generations of daughters is produced. When these are made up of cocci the clusters have been likened to tiny bunches of grapes. If successive cell divisions take place in the same plane the group of cells that is formed develops as a chain. A chain made up of cocci has been compared to a string of beads. These variations in size, shape and arrangement are depicted in Figure 5.1.

The functions of spores were described in Chapter 3, and are summarized in Box 5.1. Only a few human pathogens are capable of producing these structures (Table 5.1). Spores are typically found among microbes whose normal home is the soil, where there is a need to survive periods of drought that may last for years.

Electron microscope studies of bacteria have allowed considerable detail to be added to this simple picture. The principal anatomical features are shown in Figure 5.2 and the functions of the parts illustrated are described in Box 5.1. A bacterium is a tiny cell bounded by a cell membrane. This is protected and supported on its outside by a rigid cell wall. The cell wall is a complex and variable structure, but a feature common to all bacteria (with rare exceptions) is the presence in it of a polymer called **peptidoglycan**. This substance provides the cell with strength and rigidity, and is unique to bacteria. The pressure inside bacteria is high so that each of them resembes an inflated football. Just as a football will collapse if punctured, a bacterium with sufficient damage to its cell wall will split open and die. Peptidoglycan

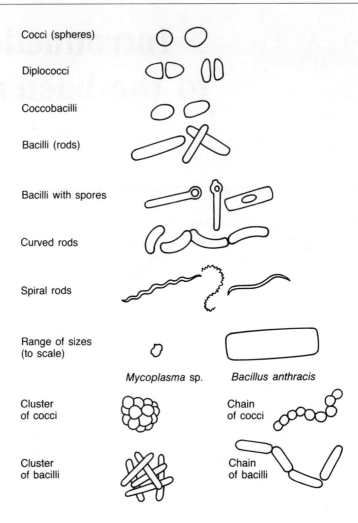

Figure 5.1 The shapes, sizes and arrangements of bacteria. (See Box 5.1 for the functions of the spores.)

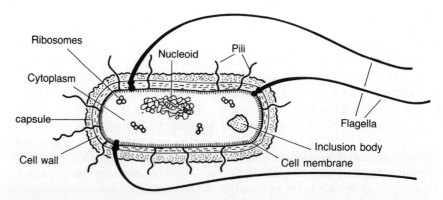

Figure 5.2 Diagrammatic representation of a composite bacterium. Not all bacteria possess all the structures shown. (See Box 5.1 for more detail and the functions of the different parts.)

is chemically unique so it provides a target for substances that, in theory at least, are poisonous for bacteria and for nothing else. This is where penicillin attacks bacteria and it explains why viruses, fungi and protozoa, which lack peptidoglycan, are not affected by it.

Box 5.1 Functions of the parts of bacteria

(See Figures 5.1 and 5.2. Parts marked * are found in most bacteria, the others are more variable.)

Nucleoid*
Naked DNA, but equivalent to nuclei in higher cells. It carries the genetic blueprint from which present and future generations are formed.

Ribosomes*
Factories made of RNA where amino acids are assembled to make proteins.

Inclusion bodies
Visible reserves of food.

Cytoplasm*
Semi-liquid ground substance.

Cell membrane*
Encloses the cytoplasm and acts as a filter that controls what passes in and out.

Cell wall*
For strength and rigidity. It may also act for offence or defence.

Capsule
Loose, often slimy outer coat, for defence or offence.

Flagellum
Single or multiple, used for locomotion.

Pili
Multiple hair-like projections for adhesion to surfaces or to other bacteria in conjugation.

Spores
Structures resistant to drying and physical or chemical attack. Can survive in conditions lethal to ordinary bacteria. When things improve they germinate and vegetative bacteria re-emerge.

The biochemistry of bacterial metabolism is a large subject. For the present purpose it is sufficient to note that it is concerned with the manufacture and assembly of the building blocks from which bacteria are constructed, and the production of the weapons they use for offence and defence. These activities consume energy, so this must be provided as well. One other feature has to be mentioned. An important waste product of metabolism is hydrogen. On its own this is an active and troublesome element and to get rid of it most living things combine it with oxygen, to make water. Many forms of life can only get rid of the hydrogen they produce in this way so they die if deprived of oxygen. Some bacteria, the **obligate aerobes**, fall into this metabolic pattern. Others are more flexible and can use a variety of chemicals in addition to oxygen as hydrogen acceptors. A few have developed these alternatives to the point that oxygen itself has become a poison and these obligate anaerobes can only grow in the absence of it. Between the two obligate extremes lie the great mass of bacteria, the **facultative anaerobes**. Most of these prefer to grow in aerobic conditions but if necessary they can manage in the absence of oxygen, usually with reduced efficiency.

5.2 STAINING BACTERIA

Early bacteriologists stained bacteria with dyes to make them easy to see under the microscope. In 1884 a Danish scientist, Christian Gram, altered an earlier method and his modification has been called Gram's stain ever since. When applied it divides bacteria into **Gram-positive** and **Gram-negative** varieties and identifies a third group that does not react with the stain. A thin smear of material thought to contain bacteria is allowed to dry on a microscope slide and heat is applied to fix it. A strong solution of the dye crystal violet is then poured onto the slide. This stains most bacteria dark blue. Iodine is added to fix the dye more firmly and the smear is washed briefly with a solvent, either alcohol or the more powerful acetone. Finally a red dye is added. If the solvent fails to remove the first dye the bacteria are Gram-positive. At the end of the process they appear purple when the smear is examined under a microscope. In the case of Gram-negative organisms the crystal violet stain is washed out by the solvent, and the once more colourless bacteria are stained red in the final stage.

Gram's stain is only reliable when applied to young bacteria. As they get old and begin to die some that are really Gram-positive become Gram-negative. The Gram's stain reactions of the more important bacterial genera are summarized in Table 5.1. Now that a good deal more is known about bacterial cell walls it is clear that there are fundamental differences between the Gram-positive and Gram-negative varieties, but it is still not known why they stain as they do.

Table 5.1 The major bacterial genera classified according to their reactions to Gram's stain, and their shapes

	Cocci	Bacilli
GRAM POSITIVE	Enterococcus Staphylococcus Streptococcus	Bacillus[1] Clostridium[1] Corynebacterium Listeria
GRAM NEGATIVE	Branhamella Neisseria	Bordetella, Brucella Enterobacteria[2] Haemophilus Pseudomonas
DO NOT STAIN	Mycoplasma Rickettsia	Leptospira Mycobacterium Treponema

[1]Capable of forming spores
[2]Includes among others the genera Escherichia, Citrobacter, Salmonella, Shigella, Klebsiella, Enterobacter, Seratia, Proteus, and Vibrio

Mycobacteria, the mycoplasmas, the rickettsias and some of the spiral bacteria do not stain by Gram's method and for them other stains or techniques must be used. The mycobacteria are made visible by a method that also helps to identify them. A hot solution of a strong red dye is applied to a smear prepared on a slide. This stains the smear and all in it a uniform, deep red. The slide is then put into a strong solution of an acid. This extracts the red dye from everything, including any bacteria that may be present, with the exception of the mycobacteria. When this stage is complete the smear is stained with a pale blue or green dye to make a pleasant background against which any red mycobacteria are easily seen. The technique is called the Ziehl-Neelsen (ZN) or **acid-fast** staining method, and because mycobacteria retain the red dye they are called **acid-fast bacilli (AFBs)**. The intensity of the decolorization process is important. If made less powerful certain other types of bacteria begin to retain the red dye, so are identified as 'weakly acid-fast', to distinguish them from those that are completely non-acid-fast.

5.3 BACTERIAL GENETICS

As outlined in Chapter 2, illustrated in Figure 5.2, and described in Box 5.1, bacterial nuclear material is not confined within a nuclear membrane, so bacteria lack a true nucleus. Instead, bacterial DNA floats free within the cytoplasm of the cell, as what is called a **nucleoid**. This distinguishes bacteria from other microbes, and from the higher forms of life.

Bacteria have made a fundamental contribution to the study of heredity and the science of genetics. Bacteria are small, breed rapidly and are easy to handle in large numbers. They can be used in genetic experiments that are cheap to perform with results that are available in a few days. The time taken for a newly-formed bacterial cell to grow and divide into two is the **generation time**. This varies between bacteria and also depends on the temperature and the availability of appropriate nourishment. Under ideal conditions the generation time for the rapidly-growing *Escherichia coli* may be only 20 minutes, while for the slow-growing *Mycobacterium tuberculosis* it is as long as 24 hours. A single fast-growing bacterium has become two in 20 minutes, eight after one hour, 64 after two, and the total has passed the two million mark in seven hours. In 24 hours a geneticist can study a number of living things equivalent to the human population of a whole country, all in a single test-tube. Compare this with the labour of a genetic study among humans with a generation time of, say, 25 years!

The DNA that makes up the genetic blueprint of living things may be altered in two ways, either by mutation or by recombination. A **mutation** is the result of a force that, applied from outside an organism, changes the structure of its DNA. The various ionizing radiations are some of the many physical and chemical forces that have this **mutagenic** effect. If the damage is severe the organism dies and the effect is not inherited. To be inherited the damage must not be lethal, though if the mutation is disadvantageous to the organism it will soon disappear. Occasionally DNA that has been altered by a mutation provides information that allows an organism to perform some task in a new and useful way. If this gives it an advantage over its neighbours the mutation is beneficial. Beneficial mutations are rare compared with those that are lethal or disadvantageous. The progress of evolution is ascribed to infinite numbers of small, random, beneficial mutations, over millions of years.

Recombination involves the union of DNA from two separate organisms be they people, animals, plants or bacteria. The transfer from one to the other may be of a large amount of DNA as happens when humans, animals or plants reproduce to form new individuals. In the case of bacteria smaller amounts may be transferred between two of them to alter the recipient, though without an increase in their number. Recombinant transfers are of pre-formed functional DNA, so their effect is much more likely to be inherited. Recombinations mix up existing genetic information while mutations introduce new genetic ideas.

Bacteria can exchange fragments of DNA between themselves in no less than four distinct ways. As genetic engineers the human race has been left far behind! One of the methods of recombination is called **conjugation**. In this process structures known as **plasmids** are transferred form one bacterium to another. Plasmids are made of DNA, and they exist inside many bacteria in a semi-autonomous way, acting like small accessory nucleoids. The information they carry may contribute little to the daily life of the bacteria that contain them, but in hard times they may determine the difference between survival and death. Conjugation involves a temporary union between individual cells that allows the transfer of plasmid DNA, so the process is of a sexual nature (Figure 5.3).

The DNA carried by plasmids may or may not be duplicated in some part of the main nucleoid of the host bacterium. In some circumstances fragments of DNA (called **transposons**) can move genes from a plasmid to the nucleoid, and back again. Alternatively the plasmid DNA may be partly or wholly distinct from anything present in the host nucleoid. The extra genetic information carried by the plasmid may be a blueprint for one or more new functions. These may help the bacterium to compete for nutrients or provide extra offensive or defensive capabilities. From the medical standpoint the fact that these additional functions can be resistant to one or more antimicrobial drugs is of particular importance. The significance of this knowledge increased when it was discovered that although conjugation is more common between bacteria of the same species, it can also take place between bacteria of different species. In this way different kinds of bacteria teach each other about antimicrobial resistance.

Plasmids reproduce themselves and copies are usually found in each daughter cell when a parent divides. In this way plasmids multiply and spread among a population of bacteria, though rather slowly. When plasmids pass for one cell to another during conjugation they spread very much more rapidly (Figure 5.3). A single plasmid donor can mate with several recipients in quick succession. Each recipient in turn becomes a new donor so in a very short time a whole population of bacteria can acquire some new ability.

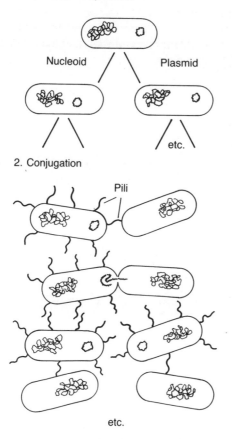

Figure 5.3 How plasmids spread among bacteria (1), slowly, as they multiply at the same speed as their hosts and (2), rapidly by conjugation when a donor passes a copy of its plasmid to a recipient, which in turn becomes another donor.

5.4 CULTURE AND IDENTIFICATION

The greatest achievement of Pasteur and Koch was to make microbes multiply in the laboratory in a controlled way in or on artificial **culture media**. Early workers used liquid culture media, but solid media have many advantages. At first gelatin was added to the liquids to solidify them. Unfortunately gelatin itself is a liquid at the temperature at which most human bacteria grow best (37°C) so this advance was of limited use to medical bacteriologists. The wife of an early bacteriologist suggested the use of agar, an extract of seaweed with a higher melting point, otherwise used for culinary purposes. This substance has been employed to solidify culture media ever since.

A single bacterium or a small clump of them begin to multiply when placed on the surface of a suitable solid medium kept at an appropriate temperature in an **incubator**. The temperature varies according to the requirements of the bacteria to be isolated, but in medical work something approaching

37°C is usual. After some time, commonly about 18 hours, the heap of bacteria that have developed is large enough to be seen by the naked eye as a bacterial colony. Colonies of the same bacterium tend to look similar when grown on the same medium, and their appearances can be used to distinguish between different species (Figure 5.4).

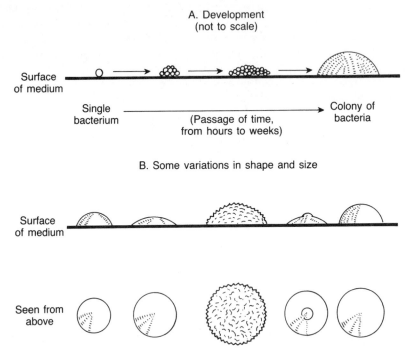

Figure 5.4 How bacterial colonies develop on solid culture media, and what some of them look like.

Certain additions can be made to culture media to provide further information about the identity of bacteria grown on them. Some bacteria produce enzymes or toxins that attack human or animal red blood cells in quite different ways, with changes that are visible to the naked eye. This is one reason for the popularity of 'blood agar' as a culture medium. Another trick is to add some particular bacterial foodstuff to the agar, together with an indicator that will change colour in the presence of acid. Bacteria often produce acids as by-products when they metabolize their food. On these special media colonies of bacteria that produce acid from the added foodstuff develop a distinctive colour. The milk sugar, lactose, is used in this way to divide bacteria into acid-producing 'lactose fermenters' and the 'non-lactose fermenters' that do not utilize the sugar at all, or do not produce acid from it. This separation is important among some Gram-negative bacilli. These 'biochemical' tests, in which bacteria are provided with a variety of potential foodstuffs ('substrates') to see which they can utilize, form the basis of a good deal of bacterial classification. When a laboratory reports the isolation of a particular bacterial species from a specimen they may well have employed some of these tests to identify it.

Another manoeuvre is used to separate one kind of bacterium from a mixture of several different varieties. Something is added to the medium to inhibit the growth of types that are not wanted so the variety sought can multiply without competition. Bacteria that live in the gut are accustomed to the presence of bile so grow happily on a culture medium that contains it, on which those that come from other places grow poorly, if at all. Many other chemicals, including antimicrobial drugs, are used in this way to produce a range of **selective media** that are a great help to bacteriologists.

The characteristics described so far play an important part in the laboratory diagnosis of infection. A specimen from a patient examined by these methods will show if bacteria are present, and give early clues to their identity and likely pathogenicity. Gram-positive cocci arranged in clusters (staphylococci) can already be distinguished from Gram-positive cocci arranged in chains (streptococci). Staphylococci that produce yellow or golden colonies are likely to be of the more pathogenic *Staphylococcus aureus*, or if white they may be of the less pathogenic *Staph. epidermidis*, though this important distinction requires to be confirmed by a more precise test. Streptococci may be separated into more pathogenic haemolytic varieties that tend to destroy or alter red blood cells added to agar and less pathogenic ones that do not. Gram-positive bacilli are readily distinguished from those that are Gram-negative, and so on. Experienced workers in bacteriology laboratories usually have some idea of the identity of the microbes they have isolated at an early stage, though further tests are often necessary to reach a final classification.

In the case of very common organisms such as the staphylococci or some salmonellas more detailed identification may be required for epidemiological purposes. It may be necessary to trace the source of a group of apparently similar infections, say in an outbreak of staphylococcal wound infection or of food poisoning. A detailed epidemiological investigation takes a great deal of time and effort and may have important consequences. Before too much is done it is useful to know that the bacteria isolated from the various patients and any other sources involved are indistinguishable, that is they are as nearly 'the same' as can be determined. If they are not, the incident is an accidental accumulation of a number of similar but not identical cases, and there is no outbreak to investigate. A variety of **typing methods** have been devised that can subdivide individual bacterial species into smaller groups. These tests are usually performed in specialized reference laboratories, where the necessary skills and scarce reagents can be concentrated.

Depending on the technique used the results of these subdivisions may be spoken of, for example, as bacteriophage types, serotypes or ribotypes. In the case of *Staph. aureus* a report may be received that identifies it as (bacterio)phage type 80/81, or some other collection of numbers. These indicate which of a series of some 30 differently numbered bacterial viruses (Chapter 16) can attack the staphylococcus that has been isolated. When the numbers allocated to each of a series of staphylococci are very similar or are the same, they are regarded as indistinguishable and they may all have come from a single epidemiological episode.

The role of the laboratory in the diagnosis of infections is discussed further in Chapter 29.

FURTHER READING

See Appendix B.

The pyogenic cocci 6
(Staphylococcus, Streptococcus, Neisseria)

6.1 INTRODUCTION

The production of pus is a consequence of many kinds of infection. Pus consists of tissue fluid modified by inflammation (Chapter 2), so it contains an excess of neutrophil polymorphonuclear leucocytes (PML) plus other inflammatory cells, and plasma proteins. Usually also present are microbes both alive and dead, and perhaps necrotic tissue. The constituents vary in quantity and quality so pus may be anything from a watery fluid to a semi-solid. It is usually coloured white or yellow. Other less usual colours may give a clue to the identity of the microbe responsible for it. Pus is often stained with blood, and if there are mucus-producing cells in the neighbourhood, it may contain mucus as well. Pus is diluted if it is produced inside, or exudes into, a fluid-filled body cavity. The presence of pus in cerebro-spinal fluid (CSF) or urine may cause a cloudiness in an otherwise clear liquid. In the case of faeces pus is invisible to the naked eye unless it is accompanied by tell-tale blood or mucus.

The presence of PML in pus is so characteristic that they are often called **pus cells**. It is normal to find a few PML in any body fluid but when numerous pus cells are seen under the microscope there is a strong presumption that inflammation is the cause. The same applies to an excess of protein. In laboratory tests on CSF and urine the number of pus cells and the concentration of protein can indicate the presence of inflammation, and so of infection. These signs may be present before the condition has progressed to make the diagnosis obvious, at a time when, in the case of CSF in particular, treatment may forestall irreparable damage. When the reports of tests are interpreted it must be remembered that inflammation can result from physical or chemical damage as well as from an infection (Chapter 29).

Although pus is characteristic of infections caused by many microbes, species of three genera of cocci are particularly noted for the production of it. The three species, labelled the **pyogenic cocci**, are *Staphylococcus aureus*, *Streptococcus pyogenes* and *Neisseria gonorrhoeae*. The first two of these bacteria are infamous as causes of infections in pre-Listerian days, when all surgical wounds became septic (Box 6.1). The pathogenic neisseria produce pus in the genital tract or over the surface of the brain in gonorrhoea and meningitis. The genera to which these pyogenic cocci belong are conveniently described together.

Box 6.1 Laudable pus

When a post-operative wound began to exude a watery blood-stained pus in the days of pre-Listerian surgery (Box 4.3) the surgeon would draw the patient's relatives aside and prepare them for the worst. A thick creamy exudate was greeted with relief. It was called 'laudable pus' because the patient had a better chance of survival. It is now known that watery pus is characteristic of the much more dangerous streptococcal infections (or gas gangrene) while creamy pus is produced by the less pathogenic staphylococci.

6.2 THE STAPHYLOCOCCI

Staphylococci are members of the normal or commensal flora of some parts of the body surface of many members of the animal kingdom. Some of them are potential pathogens and are responsible for a variety of important human and animal diseases. The principal human pathogen within the group is *Staphylococcus aureus*, though some other staphylococci also have pathogenic potential. They share a well-marked ability to develop resistance to the antimicrobial drugs used to treat the infections they cause.

The staphylococci are Gram-positive cocci, arranged in grape-like clusters. When they grow on laboratory media they develop into pigmented colonies that are commonly white or shades of yellow. They often carry bacterial viruses (**bacteriophages**) and possess plasmids. The processes of recombination allow staphylococci to exchange fragments of DNA that may transmit virulence factors and resistance to antimicrobial drugs. Obligate anaerobes that resemble staphylocci are called peptococci. They may participate in mixed anaerobic infections.

6.2.1 *Staphylococcus aureus*

(a) Pathogenesis

Human staphylococci comprise a group of potential pathogens, and commensals. *Staph. aureus* is an important cause of disease, but up to 50% of the population may carry it from time to time at some body site, most commonly just inside the nostrils, without any evidence of infection. Colonies of *Staph. aureus* often have a golden colour (Latin *Aureus* = gold). This species is distinguished from other staphylococci absolutely by the production of the enzyme coagulase ('coagulase positive staphylococci'), and variably by the production of some other virulence factors. These include toxins able to kill phagocytes (leucocidin), or cause vomiting, toxic shock syndrome or epidermonecrolysis (see below), and of enzymes able to break down various large molecules found in the tissues (fibrinolysin, hyaluronidase, lipase, deoxyribonuclease, etc.). They may also produce haemolysins. A particular feature of *Staph. aureus* is the elaboration of a cell wall component called 'protein A'. This has the property of binding with some immunoglobulin molecules, to render them inactive. The structure of the cell wall of *Staph. aureus* is illustrated in Figure 6.1.

Staphylococcal infections have a tendency to remain localized. Coagulase changes soluble fibrinogen found in blood and body fluids into insoluble fibrin, the basis of a blood clot. This contributes to the localization of infection and so to the type of pathology produced by *Staph. aureus*. In the laboratory the **coagulase test** examines the ability of a staphylococcus to cause plasma to clot. In an infection the fibrin forms a wall that protects the invading staphylococci from the body's natural defences as well as from any antimicrobial drugs used to attack it.

If *Staph. aureus* penetrates into the dermis, perhaps through a hair follicle, sweat gland or an abrasion, it can produce a furuncle (boil), stye, carbuncle or abscess, depending on the site concerned and the ability of the patient to respond more or less vigorously to the invasion. In the epidermis it produces staphylococcal (as distinct from streptococcal) impetigo. From these primary sites of infection *Staph. aureus* may leak into the bloodstream, and so spread to produce abscesses or sepsis in other parts of the body such as the lungs, pleura, endocardium, meninges, muscles, bones, joints and kidneys. *Staph. aureus* is the most common cause of surgical wound infections and osteomyelitis, and is frequently found in burns. Sometimes a large-scale invasion of the blood develops as a septicaemia. Diabetics are particularly susceptible to staphylococcal infections.

When they possess the necessary genetic information *Staph. aureus* can make exotoxins that cause various diseases. Examples are staphylococcal food poisoning, caused by eating improperly prepared and stored food in which staphylococci have multiplied to form a heat-resistant enterotoxin, the 'scalded skin syndrome' (staphylococcal epidermonecrolysis), and toxic shock syndrome. The latter is marked by fever, a severe fall in blood pressure and a scarlet fever-like rash. This syndrome was first described in epidemic form in young women using a particular brand of long-life intra-vaginal menstrual tampon (no longer on the market) that encouraged the growth of *Staph. aureus*. Scalded skin syndrome is the extreme example of the action of a toxin called exfoliatin, which causes the layers of the epidermis to separate, with the collection of clear fluid in the space formed between them. The appearance of blisters in an area of impetigo is the simplest manifestation of its action, while the shedding of large sheets of epidermis perhaps involving most of the trunk of a child, the most severe.

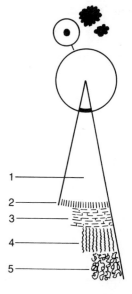

Figure 6.1 Clusters of *Staphylococcus aureus*, and the structure of the cell walls. 1. cytoplasm 2. cytoplasmic membrane 3. cell wall peptidoglycan 4. protein layer, including protein A. This can block the action of antibodies 5. capsule.

6.2.2 Other staphylococci

The remaining staphylococci are coagulase negative, and most are commensals that colonize the skin. Two of them (*Staph. epidermidis* and *Staph. saprophyticus*) are also potential pathogens. The former may cause infections in patients who have surgical implants, particularly when these are made of plastic, or who are subjected to intravenous or urinary catheterization. *Staph. saprophyticus* is a common cause of urinary tract infections in young, sexually active women.

(a) Diagnosis

A precise diagnosis of staphylococcal infection is made when the microbe is isolated by culture from material taken from a local lesion, or from the blood in more general infections. This allows it be identified, and its sensitivity to antimicrobial agents to be determined. Specimens are collected from superficial or discharging lesions with a swab, or with a syringe and needle from such deep sites as an empyema or abscess, or from blood or CSF. A common error is to send a swab when a more generous specimen of pus is readily available.

(b) Treatment

In many cases the use of antimicrobial drugs is secondary to the primary, usually surgical, treatment of staphylococcal infections. Antimicrobials do not readily penetrate into staphylococcal abscesses, so these need to be drained by incision or by some other means. Originally *Staph. aureus* was very sensitive to ordinary penicillin, but unwise overuse of this antimicrobial in the 1940s and 1950s resulted in the appearance of a variant that produced an enzyme called 'beta-lactamase' (originally penicillinase). This attacks and destroys penicillin before the antimicrobial can reach and harm the microbe that produces the enzyme. Beta-lactamase is actively secreted by staphylococci equipped with a plasmid that carries the DNA instructions for its manufacture. Resistance due to the production of this enzyme is now so common that it is dangerous to assume that a staphylococcus is sensitive to ordinary penicillin until a laboratory test has shown that it is so. Penicillins and cephalosporins can be altered chemically to resist the beta-lactamases produced by staphylococci and many other bacteria. Methicillin and its newer derivatives cloxacillin and flucloaxacillin have been modified in this way, so can be used to treat infections with staphylococci that produce beta-lactamases.

Resistance to an antimicrobial drug may also be due to a structural change in the molecule that is its target. When this happens there is nothing in the microbe for the antimicrobial to attack. Some staphylococci have changed the molecule that is the normal target for methicillin. They are no longer sensitive to this drug or to its relatives. The result is the methicillin-resistant *Staph. aureus* (MRSA) that is not only resistant to all the penicillins and cephalosporins, but often to tetracycline, erythromycin, gentamicin and chloramphenicol as well. The more expensive antimicrobial vancomycin may be the only drug suitable for the treatment of serious MRSA infections.

(c) Epidemiology and control

Most staphylococcal carriage is symptomless so although *Staph. aureus* can sometimes act as an important pathogen it is not highly pathogenic. The way in which *Staph. aureus* is transmitted can be imagined by noting the frequency with which people touch their noses, and each other. The curiosity is not that so many people carry it, but rather that a significant number do not.

Because it is so widely distributed as a part of the normal human bacterial flora little can be done to control the general mass of staphylococcal disease.

Staphylococci are a common cause of infections in hospitals, so special, though largely unavailing, efforts may be made to control the spread of it. The appearance of a case of infection with MRSA is likely to be the signal for the introduction of particularly stringent measures. Control ought to be directed towards the rational use of antimicrobials to reduce the pressure that encourages the emergence of these more resistant mutants.

6.3   THE STREPTOCOCCI

The genus *Streptococcus* comprises a large and important group of bacterial species that are responsible for much human and animal morbidity and mortality. All streptococci share three basic characteristics: they are cocci, they are Gram-positive and they arrange themselves as diplococci or in chains (Figures 5.1, 6.2 and 6.3). They are subdivided into a number of species and varieties according to which of a number of additional properties they possess. Many permutations are possible, so the classification of the strepto- cocci is complex and may be confusing. The matter is not just academic, because the additional properties determine the pathogenicity and distribu- tion of each of the types.

Initially the classification was based on easily recognized changes in the medium when streptococci were grown in the laboratory. Unfortunately there is only a loose relationship between these changes and the types of infection caused by the microbes that produce them. When streptococci grow on nutrient agar to which red blood cells have been added the colonies that develop are surrounded by one of three distinct appearances, as follows:

1. Complete destruction (**lysis**) of the red cells, with clearing of the medium so it becomes transparent. This is called **beta haemolysis**, and it is caused by **beta-haemolytic streptococci**. The toxins responsible are called **haemolysins**.
2. Partial lysis of the red cells, with incomplete clearing of the medium accompanied by a green discoloration round the colonies. This is the **alpha haemolysis** that marks the 'viridans streptococci'.
3. When there is no change in the medium the bacteria are called **non- haemolytic streptococci**.

Excluding for the present the pneumococci, the cell walls of strepto- cocci contain proteins and carbohydrates (Figure 6.2). They are made according to instructions carried by their DNA so all the offspring of a particular streptococcus possess identical molecules of these substances. Some are antigenic and when this is so antibodies appear as the result of a natural infection, or they can be prepared artificially in animals. These antibodies are used in the laboratory to recognize the presence of the same molecules in a newly isolated streptococcus. Many different proteins and carbohydrates have been identified in the cell walls of strepto- cocci, but an individual streptococcus has only one variety of each major type. Streptococci are classified according to which of these molecules they possess.

A preliminary separation of the streptococci is based on over 20 antigenically distinct carbohydrates that are found in their cell walls (Figure 6.2). These define 20-plus classes called **Lancefield groups** after the American Rebecca Lancefield, who discovered them. They are named with the letters of the alphabet. Streptococci of Lancefield group A include the major human pathogens that are usually, but not always, strongly beta-haemolytic. Members of some of the other Lancefield groups may also act as human pathogens though they are primarily pathogens of horses, cattle, pigs or other animals. Some of these are alpha- or non-haemolytic, rather than beta-haemolytic. Many streptococci possess pili by which they adhere to the surfaces of the mucous membranes they colonize or infect.

6.3.1 Group A Streptococci

(a) Pathogenesis

The human pathogens of Lancefield group A are called *Strep. pyogenes*. They have many properties, some clearly associated with pathogenicity. These include the production of exotoxins, one or more cell wall proteins, and some enzymes. Pili are accessory virulence factors that keep streptococci together as they multiply so the effects of their toxins and enzymes are concentrated and amplified.

The enzymes include streptokinase. This dissolves any fibrin that forms round an infection so contributing to the way streptococcal infections spread through the tissues. The enzyme has been used therapeutically to dissolve blood clots. Another is streptodornase which breaks down DNA into small fragments. Liquids that contain the very long strands of DNA are very viscous, and there is a lot of DNA in pus. When DNA is dissolved by streptodornase the pus becomes watery, a characteristic of streptococcal infections. A commercially available mixture of streptokinase and streptodornase may be used prior to grafting to digest and loosen the sticky sloughs on the surfaces of burns or ulcers that consist of fibrin, bacteria, pus cells and dead tissue. Streptococci also produce the enzymes hyaluronidase, leucocidin, and various proteinases.

An erythrogenic exotoxin is responsible for the rash of scarlet fever. The genetic information required to make this toxin is carried as a part of the DNA of a bacteriophage. Before a streptococcus can cause scarlet fever it must itself suffer from a virus infection! Group A streptococci also produce the haemolysin streptolysin 'O' (O = oxygen sensitive). This is antigenic, so antibodies to it develop in the course of a streptococcal infection. The presence of a large amount (a 'high titre', Chapter 29) of anti-streptolysin O (ASO) is taken to indicate that a patient is suffering or has recently suffered from, such an infection.

One of the cell wall proteins of group A streptococci is protein M (Figure 6.2). This protects the virulent streptococci that possess it from phagocytosis, so it is a very important virulence factor. When anti-M antibody develops

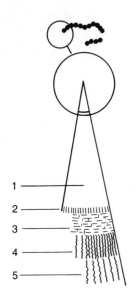

Figure 6.2 Chains of *Streptococcus pyogenes*, with the structure of their cell walls. 1. cytoplasm 2. cytoplasm membrane 3. cell wall peptidoglycan 4. Lancefield carbohydrates, over 20 varieties labelled alphabetically, A–V, etc. 5. 'M' protein, over 80 types numbered 1–80+: the major virulence factors.

as the result of an infection the individual can resist further attacks by the same M-type streptococcus. Over 80 different M proteins have been discovered so far, so there are at least 80 different 'M types' of Group A streptococci (*Strep. pyogenes*) (Figure 6.2). In theory an individual might suffer 80 streptococcal infections before he or she develops immunity to all of them, but fortunately not all types circulate simultaneously.

The classical streptococcal infection is a sore throat or tonsillitis, with or without the rash of scarlet fever. Although it is possible to suffer from a succession of streptococcal infections the erythrogenic toxins produced by all 80 'M' types are very similar, so scarlet fever usually only happens once. Other diseases caused by this bacterium are erysipelas, streptococcal pyoderma (impetigo), puerperal fever (Box 6.2), streptococcal wound sepsis and septicaemia.

Box 6.2 Puerperal sepsis

In 1844 the Hungarian Ignaz Semmelweis (1818–65), was appointed to the Obstetric Department of the Vienna General Hospital. This was divided into two clinics. At the time puerperal sepsis was the cause of heavy mortality among young mothers. This is most commonly and lethally due to *Strep. pyogenes*.

Semmelweis set out to prevent it. He succeeded, but failed as a communicator. He alienated his colleagues, was ignored, and had to leave.

	Mortality %	
	Clinic A	Clinic B
Average 1840–45	10	3
Hands disinfected before entry to ward	3	3
Hands disinfected between patients	1	1

Clinic A was used to train medical students, who also worked in the post-mortem room. Clinic B was used to train midwives who did not visit the post-mortem room.

The hands of medical staff and students carried streptococci from the bodies of those who had died of puerperal sepsis to infect new patients. The midwives only spread them between individual patients.

One to five weeks after any of these infections an individual may suffer from acute glomerulonephritis (AGN) or rheumatic fever (RF). These diseases are only indirectly due to streptococci, which have often been eradicated by the body's defences by the time one of them appears. AGN results from damage to the kidney by antigen-antibody complexes (Chapter 2). In RF, damage to heart muscles and valves is thought to be due to the development of antibodies that attack these structures. It is possible that these autoantibodies develop as the result of an infection with group A streptococci that possess antigens similar to parts of the human heart. Once the heart has been damaged in this way subsequent streptococcal infections are likely to be accompanied by a reactivation of RF with further cardiac damage. For this reason patients who have had RF need to be protected from future streptococcal infections. By contrast, second attacks of AGN are very rare.

(b) Diagnosis and treatment

As always, the best information is available when the causative organism is isolated and identified. Cultures are made of material collected on a swab taken from superficial or other accessible lesions, or of blood or other body fluids. The streptococcus isolated is identified and tested for its sensitivity to antimicrobial drugs, to help with treatment. If necessary serological tests may be performed on patients' blood to detect antibody to streptolysin O or other streptococcal products. This is useful when a diagnosis of AGN or RF needs to be confirmed after the streptococcal infection that gave rise to it has resolved and the microbe itself has disappeared.

Strep. pyogenes has remained completely sensitive to ordinary penicillin, so this is still the drug of choice. Erythromycin may be used in patients who are sensitive to penicillin, though some strains are now resistant to it.

(c) Epidemiology and control

Group A streptococci pass between individuals as a result of close contact between them, or by the contamination of food, particularly milk or milk products. They quickly lose their virulence when exposed to the air (Chapter 3).

By comparison with staphylococci, colonizations with *Strep. pyogenes* are unusual. Subclinical infections are also infrequent so most streptococci are carried by individuals who are incubating, or suffering from, streptococcal infections. For this reason an outbreak of streptococcal infection in a closed or semi-closed community (a school for example) may be controlled by the isolation of cases of the infection and the detection and isolation of carriers by swabbing the throats of apparently healthy contacts. Cases and carriers are excluded from the community and treated to eradicate the streptococcus. Alternatively the whole community may be treated in a blanket fashion with an appropriate antimicrobial drug.

6.3.2 *Streptococcus pneumoniae*

The pneumococcus, *Strep. pneumoniae*, is a rather specialized streptococcus. Pneumococci form pairs (diplococci) rather than chains and each coccus resembles the head of a lance or spear (Figure 6.3). On blood agar they are alpha-haemolytic. They differ from other alpha-haemolytic streptococci because they produce an enzyme (an autolysin) that attacks their own cell walls. The action of this enzyme is greatly potentiated by surface-active agents like detergents or bile. A broth culture of pneumococci that is cloudy becomes transparent when bile is added because the bacteria literally dissolve themselves. Pneumococci are also very sensitive to a copper-containing compound called 'optochin'. These features are used in laboratories to separate pneumococci from other viridans streptococci.

(a) Pathogenesis

Pathogenic pneumococci possess a polysaccharide capsule (Figure 6.3). This comes in over 80 varieties, each of which is antigenically distinct from all the others. Any given pneumococcus has only one kind and the ability to make it is inherited so its descendants also have the same polysaccharide. These polysaccharides are used to divide the species into 80-plus pneumococcal types. Typing can be performed by watching for the appearance of the 'capsular swelling reaction' that can be seen under a microscope when a capsulated pneumococcus is mixed with its specific antibody. In fact the swelling is an illusion, and is due to the antigen-antibody reaction that makes the extended and otherwise transparent capsule become visible.

Pneumococci without their capsules have no virulence. Unencapsulated pneumococci are regularly found among the normal bacterial flora of the upper respiratory tract. They may acquire capsules following the transfer of the necessary DNA blueprint from a capsulated pneumococcus, by recombination. In an infection the capsules act to suppress phagocytosis, so they are important virulence factors. An infection with a capsulated pneumococcus is followed by the development of antibodies to the polysaccharide of which the capsule is made. The individual is then immune to a second attack by a pneumococcus of the same capsular type. In this respect the capsular polysaccharide of the pneumococcus is very similar to the M protein of group A streptococci.

Capsulated pneumococci cause classical lobar pneumonia and empyema. They are also found in bronchitis, sinusitis and otitis media. Septicaemia may complicate any of these infections, or it may appear spontaneously. *Strep. pneumoniae* is also a cause of meningitis. It is not clear if the pneumococci responsible reach the central nervous system through the bloodstream, or by direct extension into the brain from the nasal cavity. Perhaps both routes operate at different times. Pneumococci may also invade the brain after a fracture of the skull.

(b) Diagnosis, treatment and control

The diagnosis of pneumococcal infections is by culture, though the unique shape of the coccus and its diplococcal arrangement may make it possible to recognize its presence with a microscope. Patients become unusually susceptible to infections with pneumococci after splenectomy, and they are more common among the elderly, alcoholics and the immunosuppressed. As pneumococci are widely distributed among the population, control is only possible by vaccination. A vaccine is available that immunizes against 23 of the more common capsular types. The microbe is usually sensitive to penicillin, though strains resistant to it and to chloramphenicol have begun to appear in several parts of the world.

6.3.3 The other streptococci

Group A *Strep. pyogenes* and *Strep. pneumoniae* are the major human pathogens among streptococci. The other streptococci are found regularly

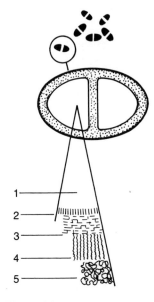

Figure 6.3 Diplococci of *Streptococcus pneumoniae* with their capsules and the structure of their cell walls. 1. cytoplasm 2. cytoplasmic membrane 3. cell wall peptiglycan 4. protein layer 5. carbohydrate capsule, 83+ numbered types: the major virulence factor.

as commensals or occasional human pathogens. They are classified as follows:

1. Streptocci of Lancefield's groups B, C, D, G, and R.
2. The viridans (alpha-haemolytic) streptococci include a group that were once lumped together as 'Strep. viridans'. They are now described as separate species.
3. A group of non-haemolytic streptococci that are also subdivided into named species.
4. The obligate anaerobic peptostreptococci that sometimes participate in mixed anaerobic infections.

Although the type of haemolysis produced by members of the first three of these groups contribute to their classification, this is a very variable characteristic. Strains may be non-haemolytic when they should be haemolytic, and the reverse.

(a)   The remaining Lancefield groups

Group B streptococci are quite often found as part of the normal flora of the adult vagina. Although their presence may be associated with a discharge, it is generally thought that they do not cause it. Group B streptococci may act as unusual opportunistic pathogens at various sites in adults of both sexes. Among neonates they can cause meningitis and septicaemia, separately or simultaneously. The streptococci that cause these infections may come from the maternal genital tract or from other patients or staff in maternity units or nurseries.

Groups C and G streptococci may be found in typical streptococcal pharyngitis and may even cause septicaemia. Many Group D streptococci are non-haemolytic and some of them have been removed to a new genus, Enterococcus (the enterococci). They are described below. Group R streptococci are important pathogens of pigs and they sometimes spread to cause meningitis in people exposed to infected animals.

(b)   Other named streptococci

This group of bacteria are found as commensals in the upper respiratory tract or, for members of Group D, in the lower part of the gastro-intestinal canal (Table 6.1). At these sites they may be involved in local sepsis such as dental abscesses or diverticulitis. Strep. mutans is strongly associated with dental caries and, with other factors, seems to be the cause of it. Dental caries is the most common human infection.

From these normal or abnormal sites streptococci regularly 'leak' into the blood stream. These leaks are associated with bodily movements and such ordinary activities as chewing or straining at stool. They may also be precipitated by dental treatment and diagnostic or therapeutic interventions in the colon, rectum or urinary tract.

In healthy people leaks of these kinds give rise to transient episodes of bacteraemia that are soon snuffed out by the defence mechanisms in the blood.

Table 6.1 The alpha- and non-haemolytic streptococci arranged by the sites and frequencies of the infections they cause. Some may appear beta-haemolytic on blood agar. Apart from the Group D streptococci they belong to the normal flora of the upper respiratory tract

Strep. spp	Bacterial Endocarditis %	Brain, abdominal or thoracic abscess %
sanguis	17	9
mitior	30	9
milleri	6	51
mutans[1]	15	0
faecalis[2]	8[3]	6[4]
bovis[2]	18[3]	9[4]
Others	6	16

1. Strongly associated with dental caries
2. Group D streptococci, normal inhabitants of the large bowel so often called entero-cocci. *Strep. faecalis* is now *Enterococcus faecalis*
3. More common in those aged over 55 years
4. Also cause urinary tract infections

Before this can happen, as a rare event, some of the bacteria may gain a foothold where there has been damage to some part of the lining of the vascular system, and begin to multiply there. If the site involved is a part of the lining of the heart, bacterial endocarditis is the result (Chapter 28). The endo-cardium may be abnormal as a result of a congenital defect, rheumatic fever, old age or surgery. As a reaction to the presence of bacteria a layer of poly-morphonuclear leucocytes and fibrin is deposited. This serves to wall them off but it also protects them from the host's immune system and from antimicrobial drugs. This is why the treatment of bacterial endocarditis is difficult, and has to be prolonged. Sequential layers of bacteria, leucocytes and fibrin form the seaweed-like 'vegetations' that characterize this disease.

When streptococci circulate in the blood and are implanted at sites other than the heart an abscess can develop. These appear most often in the abdomen, thorax or inside the skull. Different species of streptococci have preferences for the parts of the body they inhabit, the types of infections they cause and the ages of those they attack. These are outlined in Table 6.1.

(c) Treatment

With the exception of some enterococci these microbes are usually sensitive or very sensitive to ordinary penicillin. When treating severe diseases like endocarditis and meningitis, however, combination therapy is often used. Although streptococci are normally resistant to the aminoglycoside anti-microbials such as gentamicin, this is because these drugs fail to penetrate the bacterial cell walls so cannot reach the ribosomes that are their targets, which are fully sensitive to them. If gentamicin is combined with a penicillin (ampicillin is often used) the cell wall damage produced by the latter gives the former access to the ribosomes that it cannot otherwise reach.

(d) Control

Patients with endocardial damage due to any cause are exposed to an increased risk of bacterial endocarditis if for any reason they suffer a transient bacteraemia. When a leak might result from a medical or dental intervention prophylactic antimicrobials are given to protect the individual. They are used when such procedures as dental scaling or extraction, sigmoidoscopy, proctoscopy, or a cystscopy in the presence of a urinary tract infection, are performed. The drugs used may be ampicillin or amoxycillin with gentamicin or other additions if more resistant organisms such as enterococci are likely invaders, or erythromycin if there is a history of penicillin sensitivity.

6.4 THE NEISSERIAS

The neisserias are grouped with the pyogenic cocci because the major pathogens among them stimulate a brisk polymorphonuclear response when they cause infections. Other neisserias (including one that has been transferred into a new genus as *Branhamella catarrhalis*) are normal inhabitants of the respiratory tract. They may cause brochopneumonia, meningitis or endocarditis so they are also potential pathogens. Neisserias (and branhamellas) are Gram-negative cocci, with a tendency to stay together in pairs. Each coccus is roughly kidney- or bean-shaped, and the diplococci are arranged so that the flat sides of each pair are adjacent to each other (Figure 6.4).

The highly pathogenic species *Neisseria meningitidis* (the meningococcus) and *N. gonorrhoeae* (the gonococcus) are the causative organisms of meningococcal meningitis and meningococcal septicaemia, and gonorrhoea. They are capsulated and piliated (Figure 6.4). Their polysaccharide capsules probably play a part in the resistance of these organisms to intracellular killing after phagocytosis, and the pili promote adherence between them and the mucous membranes where they multiply.

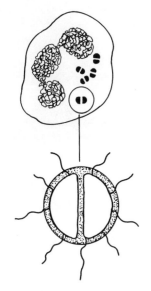

Figure 6.4 Diplococci of *Neisseria meningitidis* or *N. gonorrhoeae* located inside a pus cell, and a single diplococcus with hair-like pili for adherence and a capsule, for virulence.

6.4.1 *N. meningitidis*

(a) Pathogenesis

N. meningitidis, *Strep. pneumoniae* and *Haemophilus influenzae* are the three major causes of bacterial meningitis. The meningococci are subdivided into eight serological groups (A–D, W135, X, Y, Z) according to the antigenic structure of their capsular polysaccharides (Figure 6.4). Most disease is due to members of the groups A, B, C and W135, with one or other of these predominating at any one time in a particular geographical location. About half the cases of infection are seen in children under five years old, but no age group is spared.

The nasopharynx is the portal of entry (Figure 6.5). Meningococci that arrive there use their pili to attach to epithelial cells and multiply to produce local, usually symptomless infections. These normally resolve with the development of immunity. People with such infections are 'carriers' who can pass their meningococci on to others. In a small number of cases the bacteria spread from the pharynx into the blood to produce a bacteraemia,

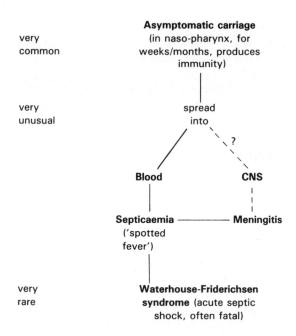

Figure 6.5 A flowchart for the development of meningococcal infections.

and by this route they can reach the meninges, the membranes that cover the brain. Sometimes the bacteraemia becomes a septicaemia with a rash ('spotted fever') before the meninges are invaded. Rarely the septicaemia develops into an acute, overwhelming infection with severe shock and intravascular coagulation before signs of meningitis can develop. This dramatic clinical entity, called the Waterhouse-Friderichsen syndrome, is often fatal. Victims of it may have a congenitally defective complement system that fails to check meningococcal multiplication.

In meningitis the meninges are acutely inflamed, with capillary thromboses and a polymorphonuclear exudate. The cerebrospinal fluid (CSF) reflects this inflammatory response, and an increase in protein and white blood cells, predominantly of the polymorphonuclear type, are found in it. The CSF is usually noticeably cloudy when a lumbar puncture is done.

(b) Laboratory diagnosis

A strong presumptive diagnosis of meningococcal meningitis is made by finding the typical organisms in a Gram-stained preparation of CSF. Under the microscope the meningococci are predominantly seen inside PML (Figure 6.4). The CSF is cultured on specially enriched media in an atmosphere that contains extra carbon dioxide. This allows the bacterium to be isolated to confirm the diagnosis and determine its sensitivity to antimicrobials. A blood culture is often positive at the same time, particularly in cases where the presentation is primarily or exclusively septicaemic. A throat swab is also likely to be positive, and this will be the only place from which the organism can be isolated in asymptomatic infections.

If antimicrobial treatment has started before a lumbar puncture is done, it may not be possible either to see or to grow the meningococci. In these circumstances the diagnosis may still be made if meningococcal capsular polysaccharide can be found in the CSF. CSF is added to tiny latex spheres coated with antibody to meningococcal polysaccharides. If sufficient capsular antigen is present the spheres stick together to form visible clumps.

(c)   Treatment

Because sulphonamides penetrate easily into the brain ('cross the blood-brain barrier') they were for a long time the drugs of choice for the treatment of meningococcal meningitis. Unfortunately over half the cases of the disease are now caused by sulphonamide-resistant strains, so a penicillin was used instead. Recently the first penicillin resistant meningococci have appeared. Third-generation cephalosporins or chloramphenicol are alternatives. Early treatment of meningitis reduces the incidence of permanent damage, and lowers the mortality. The more acute septicaemic forms of the disease progress so rapidly that treatment may be ineffective.

(d)   Epidemiology and control

The incubation period of meningococcal meningitis is between two and ten days, commonly three or four. The organism spreads in the droplets and discharges that come from the nose and throat of an infected individual, by direct contact. Carriers with asymptomatic infections are much more common than cases of the disease. A carrier rate of 25% in a population may be accompanied by no more than sporadic cases of clinical infection. In a semi-closed community (a military camp, school or prison, for example) a small epidemic of three or four cases of meningitis may be associated with a general carrier rate of 50%.

In an outbreak it is advisable to use rifampicin to treat all close (particularly family) contacts of a case. Penicillin is of no use because it fails to eradicate naso-pharyngeal carriage of meningococci, so cannot prevent its spread. All contacts should be put under surveillance to detect the first signs of illness, to allow early treatment. Vaccines may be prepared that contain the antigens of groups A, C, Y and W135. These have been used to protect individuals or communities threatened by meningococci of these groups. No satisfactory antigen has yet been prepared from meningococci of group B. Unfortunately in some parts of the world, including the UK, this group is responsible for most cases of meningocccal meningitis.

6.4.2   N. gonorrhoeae

(a)   Pathogenesis

Gonococci cause a sexually transmitted infection of mucous membranes. The genito-urinary tract is the site most commonly attacked, but the eye, anal canal and throat may also be involved. An incubation period of two to seven days is typically followed by the acute onset of a superficial suppurative inflammation. In the absence of treatment tissue invasion may follow, and fibrosis sometimes complicates healing.

In males urethritis is usual, with the production of creamy pus. The infection may spread backwards into the prostate and epididymis. Fibrosis may progress to a urethral stricture. Asymptomatic anterior urethritis is not uncommon, and the individual may then transmit the infection to others without being aware of it. Anal infections occur among male homosexuals and in others who indulge in, or are subjected to, anal intercourse. This may be asymptomatic, or it can cause discomfort and a discharge. Infections of the throat are often symptomless.

In females the infection presents as a urethritis and uterine cervicitis. This may be mild and pass unnoticed. At the next or some subsequent menstrual period, retrograde infection follows in a proportion of untreated patients. Endometritis, salpingitis and perhaps pelvic peritonitis are the result. If still untreated, damage to and fibrosis of the fallopian tubes can cause sterility or predispose to ectopic pregnancy. Chronic asymptomatic cervicitis is common. In prepubescent girls the infection develops as an acute vulvo-vaginitis, due to the susceptibility of the immature vagina. Vaginitis is rare in mature women. In infants born through an infected birth canal, gonococci can cause the severe invasive infection of the eyes called ophthalmia neonatorum. If this is not treated quickly the sight can be destroyed.

Rarely, a chronic gonococcal bacteraemia develops. This is associated with recurrent episodes of fever, arthritis (usually of the larger joints), skin lesions consisting of haemorrhagic papules and pustules, and intermittently positive blood cultures. If not recognized and treated this can go on for months or years.

(b)   Laboratory diagnosis

The characteristic Gram-negative intracellular diplococci found in gonorrhoeal pus look the same as meningococci in CSF (Figure 6.4). They may be found in smears collected from the urethra, the cervix or other infected parts. If they are seen in specimens from these sites a cautious presumptive diagnosis of gonorrhoea is made. Confirmation of the diagnosis depends on the isolation and identification of the bacterium by culture, as described for meningococci. The special media used are made selective by the addition of antimicrobials to which the gonococci are resistant, but which suppress the growth of some of the other bacteria often found in the genital tract. Many of these grow quickly so if they were not suppressed they would hide the rather slow-growing neisserias. When isolated their identity as gonococci can be confirmed, and their sensitivity to antimicrobial drugs is determined.

(c)   Treatment

Over the past 50 years the gonococcus has become progressively more resistant to ordinary penicillin, which at one time was the treatment of choice. In the early years the dose needed to produce a cure following a single dose rose steadily. Single-shot therapy is preferred by specialists in genito-urinary medicine to avoid difficulties with patient compliance. At first the increase in resistance was due to a series of mutations in the DNA of the gonococcal nucleoid. Later, in 1976, the first gonococci able to produce a beta-lactamase (penicillinase) were isolated. They had acquired a plasmid from some other bacterium that carried the DNA blueprint for this enzyme. Penicillinase-producing *N. gonorrhoeae* (PPNG) are completely resistant to penicillin, and may be resistant to other drugs as well. Resistant strains are now common and spectinomycin or a third-generation cephalosporin is used to treat the infections they cause.

(d) Epidemiology and control

The disease is spread chiefly by men and women with asymptomatic infections. Attempts are made to locate these individuals by tracing the contacts of cases when they present in clinics for the treatment of sexually transmitted diseases (STDs). This approach has not been very successful.

FURTHER READING

See Appendix B.

The aerobic and facultatively anaerobic Gram-positive rods

7

(Bacillus, Corynebacterium, Listeria, Actinomyces)

The bacterial genera dealt with in this chapter share the characteristics set out in its title. The anaerobic Gram-positive rods (clostridia) are described in Chapter 8, together with some other obligate anaerobes.

7.1 THE BACILLI

The genus *Bacillus* consists of a group of large aerobic and facultatively anaerobic Gram-positive rods that have the special property of forming spores (Box 7.1). Most species live in soil and water and on vegetation. Some are insect pathogens and some (*Bacillus cereus* in particular) may produce

---

Box 7.1 Bacillus species

There are many species of aerobic spore-bearing bacilli (ASBs), whose spores are resistant to drying, heat and disinfectants. Those that cause disease in humans, or have a medical use, are listed below, with the infections they cause.

B. anthracis
Anthrax

B. cereus (plus others)
Food poisoning; rare causes of wound and other infections.

B. stearothermophilus
Found in hot springs. The spores are very heat-resistant so are used to test autoclaves.

---

enterotoxins when they grow in food, so can cause food poisoning (Chapter 28). *B. anthracis* is the cause of anthrax, an important disease of animals that can be transmitted to humans. The spores are survival mechanisms. They are much more resistant to desiccation, heat and chemicals (including disinfectants) than the vegetative bacteria inside which they develop. Some spores can survive the action of boiling water. When favourable conditions return, spores germinate to reproduce the parent microbe in its vegetative state, and this begins to multiply once more (see Box 8.1 for the structure of a spore).

7.1.1 *Bacillus anthracis*

B. anthracis is capsulated. The capsule consists of a polypeptide which, together with other protein products of the microbe, are responsible for most of the pathology of anthrax.

(a) Pathogenesis

Anthrax is primarily a disease of herbivorous animals (sheep, cattle, horses, elephants, etc.). *B. anthracis* initially multiplies round the point of its entry into the animal body, usually in the gut. From here it spreads into the bloodstream to develop a rapidly fatal septicaemia. The blood-stained fluids that escape from the carcass contain large numbers of the bacilli and these produce spores that contaminate the ground. They remain viable for many years, and if another animal consumes them on grass or other fodder, it may be infected in turn.

In man, the pathology depends on the site of entry of the spores (Figure 7.1). Human infections develop most often as a result of handling products such as hides, hair, wool or bones from animals that have died of anthrax. If the spores gain entry through a scratch in the skin, a papule develops at the site 12–36 hours later. This quickly develops into a vesicle, then a pustule and finally into a necrotic ulcer with a black base surrounded by inflammation

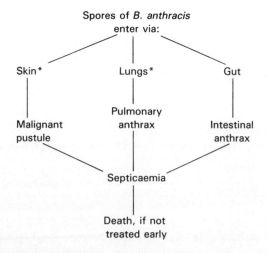

Figure 7.1 Anthrax in humans.

and oedema. If not treated in time this 'malignant pustule' may be source for the spread of *B. anthracis* into the blood, with the development of a rapidly fatal septicaemia.

If the spores are inhaled, a haemorrhagic pneumoia develops and the bacilli rapidly spread into the blood to cause septicaemia. This form of anthrax is called 'woolsorters' disease' because people employed in teasing out and separating animal wool or hair that is contaminated with anthrax spores are at special risk of it. Human infections may also follow the consumption of meat that contains *B. anthracis*.

(b) Diagnosis and treatment

Specimens of fluid or pus from local lesions or of sputum stained by Gram's method are found to contain characteristic large Gram-positive rods. These can be investigated further by culture in the laboratory. *B. anthracis* has no flagella so is non-motile, a feature that distinguishes it from most other *Bacillus* spp. Treatment is with penicillin in large doses, given before the development of septicaemia. The treatment of woolsorters' disease may fail because septicaemia develops so rapidly. Because of this, and because the spores survive very well in the air, *B. anthracis* has been suggested for use as an agent in biological (germ) warfare. The existence of a penicillin-resistant mutant makes this a particularly ugly proposal.

(c) Epidemiology and control

Control is by the disposal of carcasses of animals that have died of anthrax. They should be burnt, or buried very deeply. No part of the animal must be used for any purpose. Hides and similar products imported from areas where control is not exercised should be autoclaved, though this is not possible with bones, as heating destroys their usefulness. Workers in high-risk occupations can be vaccinated against the disease.

7.1.2 *Bacillus cereus*

B. cereus is widely distributed, and its spores are often found in dry foods such as grains and spices. If water is added to these and the food is stored at room temperature, the spores germinate and in a few hours the food contains enormous numbers of *B. cereus*. The same thing can happen if the food is cooked first and is then kept at room temperature because some of the spores survive boiling in water. As they multiply the bacteria may elaborate two different enterotoxins. One, not destroyed by further cooking, can cause vomiting; the other causes diarrhoea but is destroyed by heat, though it may be produced by bacteria in the gut itself. If contaminated food is eaten the outcome will depend on which if any of the toxins are present. The incubation period for vomiting is from one to five hours and for diarrhoea six to 16 hours. Treatment is symptomatic, and control is by the enforcement of food hygiene. Ths involves the use of adequate refrigeration to limit the amount of time food is kept at a temperature at which bacteria can multiply, and the proper storage and handling of 'leftovers', particularly rice. Some other members of the genus *Bacillus* may cause similar problems. Apart from *B. anthracis*, members of the genus are very unusual causes of tissue infections, though these are sometimes seen in infants and the immunosuppressed.

7.2   THE CORYNEBACTERIA

The corynebacteria are mainly aerobic rod-shaped Gram-positive microbes of varying length, that often arrange themselves in palisade-like bundles, or, by angulating themselves, may form patterns that have been likened to written Chinese characters. They sometimes have swollen or club-shaped ends (the word corynebacterium means 'rod-shaped club' in Greek), and special stains may reveal that they contain small inclusion bodies. Selective media have been prepared that permit the growth and recognition of corynebacteria when they would otherwise be hidden among a mass of other microbes. Some corynebacteria require fats as nutrients, and some, the proprionibacteria, are anaerobes. Most live as commensals on the skin and mucous membranes of humans and animals. A few produce powerful extracellular toxins (exotoxins) that are freely secreted into their environment.

The genus contains the important human pathogen *Corynebacterium diphtheriae*, the cause of diphtheria. The other corynebacteria are commonly lumped together as 'diphtheroids'. As has happened with other bacterial genera, several of these are now recognized as potential pathogens, particularly among the immunosuppressed. At one time diphtheroids were nearly always dismissed as contaminants whenever they appeared in cultures, including blood cultures. Because they are so widely distributed on the body surface this is often true. However it is now accepted that in some cases, species like *C. xerosis*, *C. hofmannii*, *C. pyogenes* and the more recently described 'JK' coryneform (now *C. jeikeium*) may be found acting as pathogens in patients who have been subjected to extensive surgery, particularly when transplants or implants are involved. They may appear in urinary tract infections, in bacterial endocarditis or cause septicaemia secondary to a local infection at any site. *C. minutissimum* is the cause of the skin disease erythrasma.

The anaerobic coryneform, *Proprionibacterium acnes*, produces enzymes that split the lipids found in sebaceous secretions. The fatty acids formed are thought to contribute to the inflammatory reaction in acne.

7.2.1   Diphtheria

This is an acute infection caused by toxigenic strains of *C. diphtheriae* (Box 7.2). The incubation period is two to four days. The bacterium has little invasive potential. It multiplies on skin or mucous membrane surfaces of the body, and produces the powerful exotoxin that spreads into the tissues to cause the disease. The severity of this depends on the amount of toxin that is produced, and how much of it is absorbed into the body.

On mucous surfaces, usually of the throat, the characteristic diphtheritic membrane is produced. This has been likened to a piece of wash-leather.

Box 7.2 'Sqwynancy'

This early English word was applied to an infection, almost certainly diphtheria. For centuries this was a cruel disease of nurseries. The first account of it was given in the second century AD by Arateus the Cappadocian, who described the ulceration of the throat, regurgitation of food through the nose, and suffocation. In Spain it was called 'morbus suffocans' or 'garotillo', the strangler. An epidemic of diphtheria was one of the first disasters to befall the New England colonists. The modern name for the disease (which is Greek for leather, to reflect the leathery membrane in the throat) appeared in 1826 when Pierre Bretonneau separated it from scarlet fever and croup. He may also have been the first to perform a successful tracheostomy in a case of diphtheria. Edwin Klebs discovered *C. diphtheriae* in 1883, and Friedrick Loeffler reproduced the disease in animals in 1884. The toxin was isolated by Emile Roux and Alexandre Yersin; Emil Behring developed an antitoxin from it. This was first used in a case of diphtheria on Christmas night, 1891. The first human vaccine appeared in 1913.

It lies in a shallow ulcer and round it there is a narrow zone of inflammation. On the squamous surfaces of the pharynx this slough adheres firmly to the base of the ulcer and an attempt to remove it causes bleeding. On columnar epithelium such as that found in the trachea the adhesion is less strong, and the membrane may detach spontaneously, perhaps to cause sudden choking if it blocks the airway.

The slough is made up of fibrin, leucocytes, bacteria including *C. diphtheriae* and dead epithelial cells. It is white to yellow or grey early on, but in severe cases where the membrane has spread widely it may be green or black. If the membrane forms in the larynx or trachea it, together with the associated swelling, can cause respiratory obstruction. Tracheostomy is sometimes required to overcome this, and it is needed as an emergency to relieve the acute obstruction mentioned above. On the skin, which must be damaged before *C. diphtheriae* can gain a foothold, the ulcer is similar, but the slough is usually black in colour.

Toxigenic *C. diphtheriae* are hosts to a bacterial virus (bacteriophage, Chapter 16) that inserts its own DNA into the DNA of the host bacterium. This viral DNA carries the information that allows toxin production. Non-toxigenic strains of *C. diphtheriae*, although identical to the toxigenic varieties in all other respects, do not carry the virus and are completely non-pathogenic.

The toxin acts primarily on the cardiovascular and nervous systems. The cardiovascular system is attacked in the second and third weeks of the infection. Cardiac enlargement and irregular cardiac rhythms may develop into severe, perhaps lethal circulatory collapse. Peripheral nerves are affected in the third week with paresis or paralysis of the soft palate that causes the speech to become nasal and fluids to regurgitate through the nose. In more severe cases paralysis of the eyes, pharynx, larynx, limbs and respiratory muscles follow in succeeding weeks. If the patient survives the paralyses gradually disappear.

(a) Diagnosis

The diagnosis of diphtheria should be made at the bedside. The patient may be condemned to death if this fails, or if treatment is delayed until a

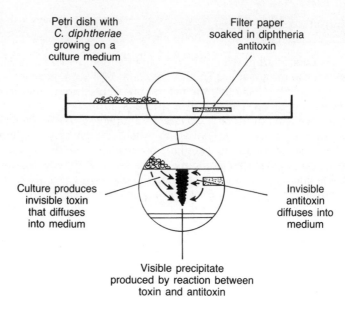

Petri dish with
C. diphtheriae
growing on a
culture medium

Filter paper
soaked in diphtheria
antitoxin

Culture produces
invisible toxin
that diffuses
into medium

Invisible
antitoxin
diffuses into
medium

Visible precipitate
produced by reaction between
toxin and antitoxin

(If no toxin is produced no precipitate is formed and the test is negative.)

Figure 7.2 The demonstration of toxin production by *C. diphtheriae* in the laboratory (the 'Elek' test).

bacteriological diagnosis has been made. Bacteriological studies are useful to confirm the clinical diagnosis and to form the basis for control measures. In the laboratory the corynebacterium is grown, identified, and its ability to produce the toxin (the proof of its virulence) is determined (Figure 7.2).

(b) Treatment

As the infection is largely an intoxication, specific therapy is with antitoxin. This must be given as soon as possible, as it is only of any use before the toxin has entered its target cells. Antibiotics are of secondary importance, but they slow down bacterial multiplication so preventing the formation of more toxin and reducing the period during which the patient may transmit the infection to others. Penicillin, ampicillin or erythromycin are used. Antitoxin is produced from the blood of immunized animals, horses in particular. When they have made enough antitoxic antibody they are bled and the antitoxin is separated from their serum. When animal products are injected into a human there is a risk of a severe allergic reactions to the foreign protein (Chapter 2), so a small test dose is given before the main one. Long-term treatment with a tetracycline or erythromycin is sometimes used in cases of acne.

(c) Epidemiology and control

Before the introduction of effective vaccines, diphtheria was a disease of children (Box 7.2). In temperate climates infection was primarily of the throat, though in tropical areas skin infections were (and in some places still are) common. Most infections were mild or subclinical. Whole populations were immunized in this way, though the benefit for the greater number was accompanied in the less fortunate by a disease with a significant mortality. Artificial immunization has removed the element of chance from the process (Chapter 31). Diphtheria toxin is made non-toxic and its antigenicity is enhanced to make the toxoid vaccine. In the triple immunizing agent for infants, this toxoid is combined with tetanus toxoid and whooping cough vaccine.

When a case of diphtheria is diagnosed, strenuous measures must be taken to prevent its spread. The patient is treated and isolated. All close contacts are identified and put under surveillance for seven days. If possible they have throat swabs taken and any carriers of *C. diphtheriae* found are also isolated and treated. Previously immunized contacts with negative swabs (or all previously immunized contacts where bacteriological examinations are not possible) are given a booster dose of vaccine, using a weaker vaccine for anyone over 10 years old, to avoid unpleasant reactions. Those with no history of immunization are given a full course of vaccine and antibiotic prophylaxis as well.

The serum of people who have had diphtheria or who have been vaccinated against it contains the diphtheria antitoxin that has been made in response. The level of this wanes in time, particularly after vaccination. The presence of the antitoxin can be detected by the Schick test. This is done by injecting a tiny dose of diphtheria toxin into the skin. After a day or two in someone who has no (or very little) antitoxin this produces a small inflammatory papule. Such a person is Schick positive, and should be vaccinated against diphtheria. If antitoxin is present the toxin is neutralized so there is no reaction. The individual is Schick negative, and immune.

C. diphtheriae var *ulcerans* is capable of producing typical diphtheria in humans, but it does not spread from person to person by direct contact as happens with the classical diphtheria bacillus. It is an animal pathogen, and it may reach humans in unpasteurized milk. The two bacteria are easily distinguished in the laboratory.

7.3 THE LISTERIA

7.3.1 *Listeria monocytogenes*

L. monocytogenes is a Gram-positive rod-shaped microbe, widely distributed among animals in whom it may cause abortion, mastitis or meningoencephalitis. Human disease due to it is more common among neonates, but adults may also be involved. *L. monocytogenes* is an aerobic Gram-positive bacillus with a characteristic tumbling motility when grown at 25°C, but not at 37°C. It has the unusual and important property of being able to multiply slowly in a refrigerator (4°C–10°C).

(a) Pathogenesis

Disease in humans due to this microbe usually presents as meningitis, preceded or accompanied by bacteraemia or septicaemia (Figure 7.3). Neonates are

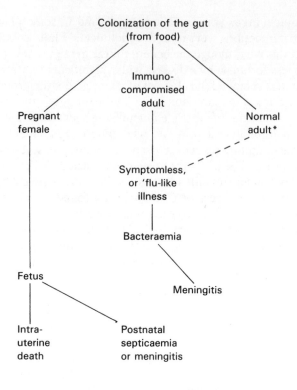

*Probably suffering from an undetected immunodeficiency

Figure 7.3 An outline of human listeriosis. Colonizations are probably quite common, but disease is rare.

particularly susceptible. They acquire the listeria from an asymptomtic infection or colonization of the maternal gastrointestinal tract, or from elsewhere. Early overwhelming septicaemia may be the result, or meningitis may appear within the first three weeks of life. Repeated abortion has been blamed on the existence of a chronic infection of the female genital tract, as may happen in animals. The evidence for this is weak and most authorities have abandoned the idea. Meningitis due to *L. monocytogenes* is seen in immunocompromised adults, and apparently normal individuals are also attacked, though very rarely The clinical picture produced is that of any acute bacterial meningitis in the age group concerned.

(b) Diagnosis

Examination of the cerebro-spinal fluid in cases of meningitis reveals the characteristic changes of a pyogenic infection, and the listeria may be seen in a smear of CSF stained by Gram's method. Culture allows the listeria to be isolated and identified, and its sensitivity to antimicrobial drugs assessed. Blood culture may be positive for *L. monocytogenes*, and in neonatal meningitis it may also be isolated form the maternal uterine cervix.

(c) Treatment

L. monocytogenes is sensitive to several antimicrobials, including ampicillin and chloramphenicol, both of which cross the blood-brain barrier so are useful in the treatment of meningitis. Unfortunately the massive use of antimicrobials to improve the economic yield of farm animals has ensured that some listeria (like many other bacteria) show resistance to these drugs.

(d) Epidemiology and control

The epidemiology of listerial infection in the human population is not yet completely understood, but the disease is almost certainly a zoonosis. The growing use of refrigeration to store milk products and other foods and the expansion of the international trade in it have introduced new routes of infection. Because *L. monocytogenes* continues to multiply at low temperatures the chilled storage of bulk foods, including soft cheeses, may convert a small number of listeria into a dangerous contamination. Because the multiplication of food-spoiling organisms that are also present is prevented by refrigeration, the products do not simultaneously 'go off' and become uneatable, as would happen at room temperature.

Food-borne listeriosis may be controlled by pasteurization or cooking prior to refrigerated storage, and this is effective if recontamination is prevented. Thorough heating before consumption is a safety measure for items that are not eaten cold, but it is necessary to remember that foods taken directly from a refrigerator to an oven take longer to reach the temperature at which they become microbiologically safe.

7.4 THE ACTINOMYCETES

The actinomycetes are filamentous bacteria that are related to the coryne-acteria becaus they are slender Gram-positive rods, to the mycobacteria by being weakly acid fast, and to the fungi by showing true branching (Figure 7.4). Most of them live freely in the soil and on vegetation. A few semi-anaerobic species of the genus *Actinomyces* are parasites in man and animals. They are found among the normal flora of the mouth and gut. From here they sometimes spread to cause actinomycosis. Some aerobic actinomycetes of the genera *Nocardia* and *Streptomyces* are potential pathogens. They cause nocardiasis and together with some true fungi have a role in the pathogenesis of mycetoma (Madura foot, maduromycosis, Chapter 24).

These Gram-positive, weakly acid-fast bacteria tend to form filaments, and may show true branching, a most unusual bacterial characteristic. With most species the filaments break up into bacillary or coccal forms. They grow rather slowly both in the laboratory and in tissues. For this reason they cause indolent, chronic infections in some of which microbial colonies develop into granules that are large enough to be seen with the naked eye. These may be coloured yellow as the 'sulphur granules' of actinomycosis, or white, red or black when the infection is due to other species.

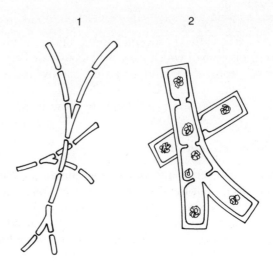

Figure 7.4 A bacterial actinomycete with branching (1), compared with a true fungus (2). Fungi are much larger and have proper nuclei.

(a) Pathogenesis

Actinomyces israelii (and less commonly *A. bovis*) are regularly found as part of the normal flora of the mouth, particularly around teeth, and in the tonsillar crypts. They can invade neighbouring tissues damaged by a tooth extraction or by another infection such as a peritonsillar abscess. Once established they cause a chronic infection called actinomycosis. When fully developed the actinomycotic lesions consist of a central area of necrosis that contains pus in which sulphur granules are found. Round this are layers of chronic inflammatory tissue and fibrosis. The whole forms an indurated lump that eventually discharges pus onto the skin through one or more channels or sinuses. These lumps enlarge slowly and rarely heal spontaneously. They develop more commonly in the neck, chest and abdominal areas.

Nocardia asteroides (and sometimes *N. brasiliensis*) live in the soil and may cause opportunistic infections called nocardiasis. This usually originates in the lungs but sometimes spreads secondarily to other organs through the blood, notably to the brain and kidneys. The primary lesion is often in an area of pulmonary damage due to some other cause, and general debility is a common predisposing factor.

Box 7.3 Mycetoma or Madura foot

Actinomycotic mycetomas are caused by the actinomycetes *Nocardia* and *Streptomyces*. The granules they produce are of several different colours. They consist of masses of filamentous bacteria, each less than 1 μm in diameter. Antimicrobial treatment is often effective.

True fungal mycetomas are caused by higher fungi. The granules they contain are made up of coarse filaments 5–10 μm in diameter (Figure 7.4). Antimicrobial treatment is ineffective and amputation is often required.

(Madura is an old name for the southern part of India).

A mycetoma is a chronic granulomatous infection usually of the foot. Skin and underlying subcutaneous tisuses are involved at first, but as the condition progresses deep fascia, muscle and bone are invaded in turn. The actinomycetes responsible live in the soil, and the disease affects those who go barefoot in tropical climates. There are two types, and it is important to distinguish between them (Box 7.3).

(b) Diagnosis

A strong presumptive diagnosis is made by microscopical examination of any granules that are found in pus. When these are crushed under a cover-slip in a drop of waer, the morphological features just described become visible. A definitive diagnosis is made by culturing the granules, and identifying the microbe that develops. In nocardiasis, granules are not formed, and a laboratory diagnosis is made by culture of clinical specimens, typically of sputum. As actinomycetes grow rather slowly, incubation must continue for several days.

(c) Treatment and control

Actinomycosis responds to long-term treatment with penicillin, though surgery may also be necessary. Nocardiasis is treated with sulphonamides, cotrimoxazole, or minocycline. The treatment of mycetoma depends on the cause. That due to the actinomycetes is susceptible to antimicrobials, perhaps accompanied by limited surgery, while the higher fungi that cause mycetoma are resistant to all the drugs at present available, so the only treatment is amputation. This is why it is important to make the correct diagnosis.

The incidence of actinomycosis can be reduced by improving dental hygiene. As an opportunistic infection little can be done to prevent nocardiasis, and mycetoma will disappear as the socio-economic status of the groups at risk improves.

FURTHER READING

See Appendix B.

8

The obligate anaerobes
(Clostridium, Bacteroides, Fusobacterium)

8.1 INTRODUCTION

This chapter introduces the obligate anaerobes of importance in human disease. Obligate anaerobes are bacteria whose growth is at least inhibited by oxygen; some of them are killed by it (Chapter 5). Of the many genera that exist in nature, Table 8.1 lists those that are well established as causes of human pathology. The more important obligate anaerobes are dealt with together because their inability to multiply or even to survive in the presence of oxygen imposes upon them similarities that outweigh their differences. The allocation of microbes to aerobic and anaerobic groups is not clear-cut because nature has experimented with intermediate forms. Some microbes have developed to operate best at oxygen levels below those found in normal air. These 'semi-aerobic and facultatively anaerobic' (or microaerophilic) bacteria include members of the genera *Actinomyces* and *Campylobacter* that are more conveniently described elsewhere (Chapters 7 and 11).

Table 8.1 Anaerobes of clinical importance

BACILLI	Gram-positive
	Clostridium
	Actinomyces[1]
	Proprionibacterium[2]
	Gram-negative
	Bacteroides
	Fusobacterium
	Campylobacter[1]
COCCI	Gram-positive
	Peptococcus[3]
	Peptostreptococcus[3]

[1] Semi-aerobic genera described in Chapters 7 and 11
[2] Chapter 7
[3] Unusual pathogens, dealt with in Chapter 6

Most of the species within the genera listed in the title of this chapter are to be found as part of the normal flora of humans and animals. Their existence indicates that significant areas of the body surface are at least somewhat anaerobic. This is particularly true of the lower part of the intestinal canal.

Table 8.2 Distribution of the anaerobic normal human flora

Normal habitat	Clostridium	Bacteroides	Fusobacterium	Cocci
Pharynx	+/−	+++	+++	++
Lower GI tract	++	+++	++	+++
Vagina	+/−	++	+	++
Skin	+/−	+/−	−	++

GI = gastrointestinal
−, absent or rare; + to +++, found with increasing frequency

So far as humans are concerned, most of these commensals are also potential pathogens. Their normal anatomical distribution is shown in Table 8.2. Two notable general facts about anaerobic infections are that they may be associated with a particularly repugnant smell, and that metronidazole is a very effective form of treatment for them.

The cultivation and identification of anaerobes in the laboratory is complicated by the need to provide special conditions for their growth. All require a lower concentration of oxygen than is found in air or in normally oxygenated tissues, though some are more tolerant of oxygen than others. In practice it is convenient to provide an incubator and a workstation in a single unit, from which virtually all the oxygen has been removed. The growth of anaerobes is often improved when they are provided with extra carbon dioxide. Specimens for anaerobic culture should not be exposed to oxygen in the air for longer than absolutely necessary. A few millilitres of pus in a sterile bottle or a syringe is much better than a swab but if one of these has to be used it is best sent to the laboratory with its tip embedded in transport medium.

8.2   THE CLOSTRIDIA

Clostridia are Gram-positive rods that form spores (Box 8.1). Spores survive conditions that cause the death of bacteria in their vegetative state. Most clostridia live in the gastrointestinal tract of humans or animals,

Box 8.1 Bacterial spores

Spores are found in aerobic and anaerobic Gram-positive bacilli of the general *Clostridium* and *Bacillus* (Chapter 7).

Each spore-bearing species tends to produce its spores in a characteristic way: they may be round or oval in shape and while still inside a bacillus they may be sited centrally (equatorial), near one end (subterminal) or right at the end (terminal). They may or may not distend the bacillus inside which they are formed.

Inside the spore is the DNA bacterial nucleoid, plus some dehydrated cytoplasm. Round this is the cell membrane and several coats, including one that is impermeable to water and to many chemicals. Spores allow the bacteria that produce them to withstand adverse conditions, drought in particular. The partial dehydration gives resistance to heat but total dehydration and death are prevented by the impermeable coat that also enables the spore to resist chemicals, including most disinfectants. Spores germinate and new bacterial cells emerge when favourable conditions return.

so they are also found in the soil. For each species, spores tend to form within the bacterial body in a characteristic way. The variations may be helpful in the recognition of a particular clostridium: *Cl. tetani* for example produces swollen terminal spores, so the complete bacterium resembles a drumstick. Clostridia cannot utilize oxygen as a hydrogen acceptor and in the presence of oxygen toxic materials appear that at least inhibit their growth, and may kill them. Clostridia are associated with the human diseases shown in Table 8.3.

Table 8.3 Human pathology due to the clostridia

Wound infections	
Gas gangrene	*Cl. perfringens*
(or less commonly)	*Cl. .novyi* *Cl. septicum* *Cl. histolyticum*
Tetanus	*C. tetani*

Enteric syndromes	
Botulism	*Cl. botulinum*
Food poisoning	*Cl. perfringens*, type A
Jejunitis (pig-bel)	*Cl. perfringens*, .Type C
Pseudomembranous colitis	*Cl. difficile*

Clostridia produce a number of highly active exotoxins that are responsible for much of the pathology they cause. Some also produce enzymes that can split tissues so encouraging the development of invasive, spreading infections. Other toxins attack specific targets such as cell membranes.

8.2.1 Gas gangrene

(a) Pathogenesis

Gas gangrene nearly always starts in a wound, usually of a limb. Within the wound there must be an anaerobic (oxygen-deficient) area. This may be muscle or other tissue damaged at the time of the injury, or a large blood-clot. Next, this anaerobic tissue must be contaminated with clostridial spores, particularly those of *Cl. perfringens*. If the oxygen tension is low enough, these will germinate, revert to their vegetative state and begin to multiply. *Cl. perfringens* is a normal inhabitant of the large bowel of humans and animals, so its spores are found almost everywhere in soil and dust as well as on the skin, particularly below the waist.

Once clostridia are established in an anaerobic wound their enzymes and toxins diffuse outwards, and more tissue is killed. This enlarges the anaerobic area into which the clostridia can spread. Gas is produced by bacterial enzymes as they attack the tissues, and this adds to the pathology as it splits muscles apart to allow the infection to extend. It also inflates bundles of muscles inside their relatively rigid fascial sheaths. When the pressure of the gas rises above the arterial pressure the blood supply to the bundle is cut off, and the whole of it becomes anaerobic. This happens more easily because the blood pressure is likely to be low as a result of the original injury and it is reduced still further as the absorption of toxins leads to septic shock. When these factors combine, gangrene can spread to a whole limb in a matter of hours, and death soon follows. A mixture of microbes is nearly always found in such cases, but one or more of the clostridia named in Table 8.3 is always present.

Gas gangrene is an infection of muscle. It must not be confused with clostridial infections of other parts of the body. The development of gas in subcutaneous connective tissues is not gas gangrene and it is not an immediate threat to life. In such cases the emergency amputation on which survival may depend in true gas gangrene is unnecessary.

(b) Diagnosis

The diagnosis of gas gangrene is made at the bedside or in the operating room. A bacteriological examination may show the presence in a wound of a bacterium capable of producing gas gangrene. This does not make the diagnosis, even when gas is present in or near it. Gas must be present in and around dead muscle for the diagnosis of gas gangrene to be made.

(c) Treatment

Treatment of gas gangrene is urgent, and consists of the radical surgical removal of all affected tissue. Penicillin and metronidazole are given. A hyperbaric (pressure) chamber like those used to treat a diver suffering from the 'bends' may help. The growth of anaerobic organisms may be inhibited if oxygen can be driven into dead tissues by the raised atmospheric pressure inside the chamber.

(d) Control

Prevention involves early thorough cleaning of dirty wounds with the removal of all dead or devitalized tissue. Where an injury has caused considerable tissue destruction the wound should not be sutured immediately but is left open for a time ('delayed primary suture', DPS). This is to reduce the possibility that an anaerobic area will develop in a closed wound in which tissues may continue to swell as inflammation develops. The rule applies to severe compound fractures or crush, missile or blast injuries. Injuries of these kinds are common in war. Some surgeons accustomed to peacetime surgery forget the need for DPS when faced with battle casualties for the first time, and unnecessary cases of gas gangrene begin to appear. Gas gangrene can also complicate amputations performed for diabetic or arteriosclerotic gangrene. Prophylactic penicillin should be given in all these cases.

8.2.2 Tetanus

(a) Pathogenesis

Cl. tetani multiplies in the gastrointestinal canal of animals, particularly large herbivores. The powerful exotoxin produced by these bacteria is destroyed by digestive enzymes, so when confined to the gut it does no harm. Spores of *Cl. tetani* are found in soil contaminated by faeces, where they survive for years. If a wound is contaminated with some of these spores, and when it also has within it a sufficiently anaerobic area, the spores germinate, multiply locally, and produce tetanus toxin. This may happen shortly after the injury, or perhaps years later when some new accident produces an anaerobic area in an old wound that contains dormant tetanus spores.

The toxin diffuses into the circulation, and so reaches the spinal cord. Tetanus toxin is a neurotoxin that acts to counteract the inhibition of the spinal stretch reflex. This is the reflex that is responsible for the 'knee jerk'. When this reflex is not suppressed voluntary muscles cannot act in a coordinated fashion because the contraction of one muscle group stimulates the contraction of the opposing group, to produce a painful spasm. In tetanus spasms appear in response to stimuli such as touch or noise. At each joint the more powerful group of muscles overcomes the weaker. The very characteristic attitude of a patient in a major tetanic spasm ('opisthotonus') depends on this. The muscles that keep the spine erect are stronger than those that bend it. In a tetanic spasm the back of the patient is strongly arched so that he or she may be poised on the back of the head and the heels, with the rest of the body off the ground. In the early stages of tetanus local groups of muscles may be involved. 'Lockjaw' is due to spasm of the powerful muscles used for chewing, and the fixed grin ('risus sardonicus') of tetanus is caused by the contraction of the muscles that control the corners of the mouth.

Tetanus most commonly appears as the result of an accidental wound, which may be trivial. A thorn from a plant embedded in the dermis is sufficient. Tetanus may also complicate surgical operations, and the umbilical stump of a newborn infant provides a very suitable place for *Cl. tetani* to multiply. Neonatal tetanus is an important cause of infant mortality in countries where traditional birth assistants are untrained and ritual mixtures that may contain tetanus spores are applied to the stump.

(b) Diagnosis, treatment and control

The diagnosis of tetanus is clinical, though if the site of the wound in which it is multiplying can be identified and is accessible, the causative bacterium can be isolated and identified in the laboratory.

Treatment is of two kinds. First is the relief of the painful muscle spasms which, if severe, soon exhaust and kill the patient. To prevent this the patient is paralysed with drugs and artificial ventilation is used to maintain respiration. Second, antitoxic antibodies are given to neutralize any toxin not already bound to nerve tissue. An antitoxin produced in humans is available. This does not carry the risk of hypersensitivity associated with animal sera.

If horse or sheep antitoxin has to be used, a small test dose is given first to detect those who are allergic to it.

If the wound in which *Cl. tetani* is multiplying can be identified, it may be explored surgically to rid the body of the dead tissue responsible for the production of the toxin. Although of limited use, antibiotics are given. A patient recovering from tetanus requires to be immunized artificially to prevent a recurrence, particularly if the source of the toxin has not been identified and eliminated. The reason for this is that the amount of toxin that can cause a severe attack of tetanus is insufficient to immunize the patient.

Tetanus is controlled and prevented by immunization, and re-immunization as required, with tetanus toxoid. Neonatal tetanus can be prevented if the mother is immunized late in pregnancy so that the antibodies she develops can pass through the placenta to protect her child after it is born.

8.2.3 Food poisoning due to *Cl. botulinum*

(a) Pathogenesis

Botulism is a rare but very lethal form of food poisoning. Botulinus toxin is produced when *Cl. botulinum* grows under sufficiently anaerobic conditions, usually in food. The spores are widely distributed in nature. There are several different antigenic types of the toxic protein of which types A, B and E may affect humans. This toxin is one of the most poisonous substances known, but it can be destroyed by boiling for 20 minutes. Because *Cl. botulinum* and its spores are found everywhere, food is easily contaminated. Fortunately the spores will not germinate unless the conditions are extremely anaerobic, and the food is alkaline. Tinned or vacuum-packed foods that are not cooked further before consumption have most often been implicated as the source of the toxin.

The toxin interrupts the transmission of nerve signals at the junctions between nerves and muscles, so it causes paralysis. Symptoms appear 18–96 hours after the toxic food has been eaten, and if untreated difficulty with swallowing and speech are quickly followed by respiratory paralysis and death. The disease in an atypical form is seen in infants following the colonization of the gut by *Cl. botulium*, and cases of botulism complicating wound infections have been reported.

Box 8.2 Great fleas and lesser fleas

So, naturalists observe, a flea
Hath smaller fleas that on him prey
And these have smaller fleas to bite'em
And so proceed *ad infinitum*.

On Poetry
Jonathan Swift (1667–1745)

The production of the erythrogenic toxin of scarlet fever (Chapter 6) and the diptheria and shigella toxins (Chapters 7 and 10) depend on the infection of the bacteria concerned with bacterial viruses (bacteriophages) that carry the appropriate DNA. The same is true of botulinus toxin. It is a curiosity that some bacteria can only cause human infections if they themselves are infected by viruses (Box 8.2).

(b) Diagnosis, treatment and control

Botulism is diagnosed clinically. In the laboratory the toxin can be identified in the serum of the patient, and in the left-overs of the food, from which *Cl. botulinum* may also be isolated. Specific treatment is with antitoxin, using a mixture of the three antigenic types that affect humans. General treatment is directed to the maintenance of respiration and other vital functions until the paralysis begins to recover spontaneously. The food industry is very aware of the dangers of botulism, and takes care to avoid it, but accidents still happen. It is very important to inform the local public health authority if botulism is suspected so that a search for the cause can begin and any suspect food be withdrawn before more people are poisoned.

8.2.4 Food poisoning due to *Cl. perfringens*

Cl. perfringens can multiply in food, especially in meat dishes kept at room temperature. Trouble starts when the numbers of heat-resistant type 'A' *Cl. perfringens* in each gram of food climb above 100 000. When these form spores in the lower gastrointestinal tract they release sufficient enterotoxin to cause acute diarrhoea and colicky abdominal pain. Vomiting is unusual. Symptoms appear 12–20 hours after eating the meal concerned, and may persist for up to three days. This is at best unpleasant and among elderly and debilitated individuals it can be lethal.

The diagnosis is made by isolating the causative clostridium from the stools of patients, and from the incriminated food. Treatment is supportive only, with intravenous fluids if diarrhoea has caused severe dehydration. Control consists of proper handling of food, remembering that small numbers of heat-resistant type 'A' *Cl. perfringens* may already be present in raw meat. The key is effective refrigeration before cooking, and, because the spores are resistant to heat, afterwards as well if the food is stored before it is consumed.

Pig-bel is a necrotizing jejunitis seen in Papua New Guinea in people who eat pig meat that contains large numbers of *Cl. perfringens* of type 'C'. This happens when there is a festival at which many pigs are slaughtered, with no refrigeration to keep the meat in good condition. The infection may be fatal, particularly among children. At one time the disease was not uncommon in northern Germany, where it was called *darmbrand* ('fire-belly').

8.2.5 Pseudomembranous colitis due to *Cl. difficile*

This often severe and potentially lethal ulceration of the colon is due to the overgrowth of *Cl. difficile* in the gut. The condition usually appears after antimicrobial drugs have decimated the normal, predominantly sensitive, bacterial flora. In the absence of competition the resistant clostridia can grow excessively and produce a toxin that causes the pathology. Treatment consists of stopping the antimicrobials that have precipitated the condition, and giving vancomycin or metronidazole (to which *Cl. difficile* is sensitive) by mouth. Pseudomembranous colitis is increasingly common among older patients in hospitals. Control is difficult because of the widespread use of antimicrobial drugs and because, unusually for hospital pathogens, the spores survive well in the environment.

8.3   *BACTEROIDES* AND *FUSOBACTERIUM*

These are Gram-negative non-sporing anaerobic bacilli. They are found as normal inhabitants of most of the anaerobic areas of the body (Table 8.2). From these sites they may spread to cause mixed anaerobic infections. For example, anaerobes in the oropharynx may become involved in sinusitis or otitis media, usually in the company of other microbes like the facultatively anaerobic enterobacteria. Such infections may extend to produce a brain abscess. These bacteria may also be found in peritonitis or a subphrenic abscess that develops after appendicitis, diverticulitis or abdominal surgery. The species most often involved in this way are *Bacteroides fragilis* and *Fusobacterium necrophorum*.

The laboratory diagnosis of these mixed infections is by the culture of pus under aerobic and anaerobic conditions. The true anaerobes have to be separated from any facultatively anaerobic companions present at the same time before they can be identified and antimicrobial sensitivity tests performed. This is technically difficult and takes time. In such cases the initial treatment must take account of the range of microbes likely to be present. Two or more antimicrobials may be needed and metronidazole is often included to deal with the anaerobes. As always, any abscess cavities should be drained as part of the treatment of the patient.

FURTHER READING

See Appendix B.

9 The mycobacteria
(Mycobacterium)

9.1 THE MYCOBACTERIA

Most mycobacteria are harmless inhabitants of soil and water, but some are important pathogens of animals and birds. In humans they cause tuberculosis and leprosy, diseases with long histories that have maimed and killed enormous numbers of people. Both are still common in certain parts of the world, and the incidence of tuberculosis has begun to rise again in more developed countries not least because of the pandemic of AIDS.

Mycobacteria are rod-shaped organisms. They cannot be stained by Gram's stain, but the structure of their cell walls indicates that if it were possible, they would be Gram-positive. To make them visible under the microscope they are treated with a red dye, basic fuchsin. This is combined with phenol, and applied hot (Chapter 5). Stained in this way they resist decolourization with a mixture of alcohol and strong acid that washes the dye out of almost everything else. Under the microscope any mycobacteria present are still bright red so they are called acid-fast bacilli (AFBs). This property is associated with the presence of a waxy material in mycobacterial cell walls. Mycobacteria grow slowly or very slowly both in tissues and in the laboratory. Faster growing species take between 48 hours and 10 days to produce visible colonies while the more pathogenic ones may take up to eight weeks for there to be sufficient growth to be seen with the naked eye.

The waxy coats of mycobacteria give them considerable resistance to chemical disinfectants and to drying. In these respects they resemble spores, but as their cytoplasm is not dehydrated they are fully sensitive to heat.

Box 9.1 Human disease caused by mycobacteria

M. tuberculosis, the human tubercle bacillus, an obligate aerobe.

M. bovis, a semi-anaerobic pathogen of cattle and other mammals; another cause of tuberculosis in humans.

M. leprae is the cause of leprosy. It has not been grown on laboratory media, though it will multiply in immunocompromised mice and in the nine-banded armadillo.

'Opportunistic mycobacteria' include *M. intracellulare*, *M. kansasii*, *M. avium*, *M. fortuitum*, *M. chelonei*, and *M. marinum*. These cause symptomless or minor infections in normal people, or act as more significant pathogens in debilitated or immunocompromised individuals. Diseases due to these less pathogenic agents are called the mycobacterioses. They can be of importance in patients with AIDS.

The temperature at which milk is pasteurized (63°C for 30 minutes, or 72°C for 15 seconds) was chosen with the need to kill *Mycobacterium bovis* in mind. The species of mycobacteria that causes human disease are listed in Box 9.1.

Mycobacterial cell walls contain antigenic proteins. There are small but antigenically significant differences between the proteins of the various mycobacterial species that may infect humans.

9.1.1 *Mycobacterium tuberculosis* and *M. bovis*

Human tuberculosis is caused by *M. tuberculosis* or *M. bovis*. No single factor or group of factors have been identified to account for the special pathogenicity of these bacteria. Their waxes and protein antigens stimulate a cellular immune response that is responsible for the way the tissues react in tuberculosis.

(a) Pathogenesis

In the non-immune host, inhaled tubercle bacteria establish themselves in the respiratory tract. The primary disease is then of the lungs and the related lymph nodes. This kind of infection is nearly always due to *M. tuberculosis*. Infection with *M. bovis* is usually acquired by drinking raw milk from cows with tuberculous mastitis. ('Raw' means milk that has not been pasteurized or sterilized.) Human infections with bovine tubercle bacilli are primarily of the gastrointestinal tract and they involve the tonsils and cervical lymph glands, or the intestine and the lymph glands in the abdomen. Cattle infected with *M. bovis* can suffer from a disease that resembles human pulmonary tuberculosis. Farmers and others who work with them may also develop lung infections caused by the bovine tubercle bacillus.

Mycobacteria are intracellular parasites that live and multiply inside phagocytic cells or in the 'giant cells' that form when several of these fuse together. The intracellular habitat favours microbial persistence, and complicates antimicrobial therapy. The non-immune host reacts to the implantation of tubercle bacilli by a local inflammatory reaction, with a monocytic and macrophage response. During this phase the patient develops cellular immunity (hypersensitivity) to the protein antigens of the microbe. Phagocytes fail to kill the bacteria and they begin to multiply as they are carried through lymphatic channels to the local lymph nodes, where most are trapped by the inflammatory reaction that develops there. There are thus two lesions, one at the site of implantation and the other in the local lymph node. The combination, called the **primary complex**, usually heals without causing significant disease. It leaves behind a fibrotic or calcified scar, the **Ghon focus**. For most of those infected this is the end of the matter. As a result of a symptomless or trivial infection they have acquired a protective cellular immunity (Box 9.2).

In a minority of individuals the disease progresses, and characteristic **tubercles** are produced. These may heal in their turn, or enlarge to form abscess cavities that contain 'caseous' pus, with the consistency of a soft cheese. As the disease progresses in the lungs the enlargement eventually

---

Box 9.2 Pathology in tuberculosis

M. tuberculosis is implanted in the lungs (by inhalation) or *M. bovis* in the gut (by ingestion). A 'primary complex' develops as a pair of lesions, one at the site of implantation and the other in the local lymph glands. This reaction produces a cell-mediated delayed hypersensitivity to the mycobacteria.

Over 90% of these lesions heal, often with calcium laid down in them so they are seen in X-rays (the 'Ghon focus'). Tubercle bacilli remain viable within these and may break out to cause relapses later.

In about 10% of cases the primary complex does not heal and the disease continues to develop. Before treatment was possible this might eventually heal, but failing this the disease became chronic and finally lethal.

---

involves a bronchus, and the caseous material with the tubercle bacilli it contains is coughed up, perhaps to infect others. Such an individual has **open tuberculosis**. The process is accompanied by the development of fibrous tissue. If the patient cannot react with sufficient vigour the lesions continue to increase in size until death follows. This may be due to the progressive destruction of healthy lung tissue, secondary infection, or from a haemorrhage when the process erodes a large artery. The disease may spread to involve the skin, bone, joints, the central nervous system, kidneys, peritoneum, pericardium and so on. Individuals whose immune response is totally inadequate develop miliary tuberculosis as the bacilli spreads rapidly and widely throughout the body.

Those who overcome their infection, with or without treatment, often continue to harbour living tubercle bacilli in their 'healed' lesions. These can cause **reactivation tuberculosis** if the delicate balance between a patient's resistance and the capacity of mycobacteria to multiply is upset in favour of the microbe (Box 9.2). This happens most often in older people who may then develop open tuberculosis with minimal signs of illness. Individuals in this condition are particularly dangerous because they produce sputum that contains large numbers of tubercle bacilli.

(b) Diagnosis

The cellular immunity that develops persists when the infection becomes dormant. The few bacilli that remain in the tissues may contribute to this. The result is a resistance to reinfection the existence of which can be detected if a small amount of the sensitizing mycobacterial protein is introduced into the skin. Originally a rather crude material was employed, called old tuberculin. This was prepared from cultures of *M. tuberculosis*, and the test was called the **tuberculin test**. Nowadays an improved 'purified protein derivative' (PPD) is used instead.

To perform the test PPD may be injected intradermally with a syringe (the Mantoux test) or it may be implanted into the dermis on needles driven into the skin by a spring specially designed 'gun' (the Heaf test). Other methods are also used. The PPD is carefully standardized to make tests comparable, and also because too much of it can cause a severe inflammatory reaction in patients with active tuberculosis. This was the mistake Koch made when he used tuberculin to treat tuberculosis (end of Box 4.3).

An individual who has not developed cellular immunity to the tubercle bacillus either has never met it before, or, much less commonly, has failed to react to the invasion so is suffering from an overwhelming miliary infection. In either case there is no reaction to the injection and the test is negative. In individuals who have developed immunity a hard, red, swollen area appears 24–48 hours after the injection. The size of this positive reaction is measured, and the result is interpreted according to standard rules. A positive test indicates that the patient has had tuberculosis at some time, or has been successfully vaccinated against it. Active disease is unlikely to be present, particularly if the reaction is weak. An unusually large reaction suggests recent or even current infection, and prompts a search for its cause. This may be the existence in the patient of early clinical or subclinical tuberculous disease, perhaps acquired from a case of open tuberculosis among friends or relations.

In the laboratory a presumptive diagnosis of tuberculosis is made by the demonstration of AFBs in the relevant specimen. In pulmonary tuberculosis this is usually sputum, and in other cases the type of specimen depends on the site of the infection. In patients with suspected pulmonary disease who produce no sputum, laryngeal swabs or gastric washings may be used instead. The findings of AFBs in a patient with symptoms and signs suggestive of tuberculosis allows treatment to be started without the wait of several weeks needed for the definitive diagnosis provided by a positive culture. Because the non- or weakly pathogenic mycobacteria cannot be distinguished from *M. tuberculosis* or *M. bovis* under the microscope, culture is necessary to confirm the diagnosis. Not only does this allow the causative organism to be identifed positively, but also its sensitivity to antimicrobical agents can be determined. Because some tubercle bacilli are resistant to one or more of the standard antituberculous drugs this can be useful information. Tubercle bacilli grow very slowly, so special techniques are used. Because mycobacteria are resistant to many chemicals, specimens can be treated in ways that kill other bacterial contaminants that would otherwise grow much more quickly. Cultures made on special media are incubated for six or eight weeks before they are declared negative, though positive results are often available rather sooner.

(c) Treatment

Tuberculosis, and attempts at its treatment, have a long history (Box 9.3). It was not until streptomycin appeared in the middle of the twentieth century that any real progress was possible. In general, antimicrobial drugs are only effective against actively growing bacteria. Mycobacteria grow slowly so treatment must be prolonged. Because they multiply inside cells, the drugs used must be able to penetrate these to reach the bacteria they contain. The long period of treatment gives the organisms an unusual opportunity to become resistant to an antimicrobial while it is being given. Resistance is less likely to develop when two or three drugs are given simultaneously, so

Box 9.3 The King's Evil

Evidence of tuberculosis has been found in an Egyptian mummy of 1000 BC. Hippocrates seems to have known about it, and Galen recognized that it was contagious.

Scrofula was tuberculosis of the lymph glands, and in France this 'King's Evil' was treated by the Royal Touch. The practice was brought to England by Edward the Confessor (1003–1066). Charles II is said to have touched nearly 100 000 during his reign, 8500 in 1682 alone. In 1864 the crowd of the sick was so dense that some were trampled to death. As a child in the reign of Queen Anne, the lexicographer Samuel Johnson was one of the last to receive the Touch, in 1712.

Robert Koch identified the tubercle bacillus in 1882, and in 1924 Albert Calmette and Camille Guérin introduced their vaccine. The discovery of streptomycin in 1943 by Selman Waksman began a new era in the treatment of the disease, now overshadowed as the tubercle bacillus has begun to develop serious resistance to many antituberculous drugs.

antituberculous agents are prescribed in combinations. These are often based on isoniazid, combined with one or more of rifampicin, streptomycin or ethambutol. In the case of resistance, drugs such as ethionamide, pyrazinamide, cycloserine and para-aminosalicylic acid are substituted. Many different combinations have been tested in an attempt to develop low-cost treatments that are more likely to secure patient compliance. The long courses needed to achieve a cure are not easily accepted by patients who feel much better after a week or two, particularly if the drugs are difficult to take, or produce unpleasant side-effects. As patients are usually treated at home this is a major practical problem.

A recent unwelcome development has been the appearance of tubercle bacilli that are simultaneously resistant to several antituberculous drugs. Antimicrobial drugs can only 'cure' infections if patients' immune systems collaborate. When immunity is deficient a cure may not be achieved. This is why some patients (cases of AIDS for example) do not respond adequately to antituberculous treatment. When such patients develop tuberculosis and are treated for it, unusually large numbers of tubercle bacilli are exposed to antituberculous drugs for exceptionally long periods. It is difficult to imagine a better way to encourage the emergence of resistant bacteria.

One consequence of this development is that health care workers may come into contact with patients suffering from open tuberculosis caused by multi-resistant bacteria. If they are not already immune they may acquire a tuberculous infection. If this develops into a progressive disease their infection may be difficult or impossible to treat with antimicrobial drugs. This has far-reaching implications, and suggests that any who are not already immune, and who are likely to be exposed in this way, should be vaccinated against the infection.

(d) Epidemiology and control

In a population where the spread of tuberculosis is not controlled, perhaps 10% of people infected with tubercle bacilli become ill. Before antituberculous treatment was available 10% of this group (that is, 1% of the infected population) might develop progressive, life-threatening illness (Box 9.2).

A single individual with open tuberculosis may infect many hundreds of susceptible neighbours. The cellular immunity that results from an infection gives some protection from further infection. This process may be mimicked by vaccination, and a whole population can be given some immunity without the morbidity and mortality that accompanies natural infections (Box 31.2). The vaccine is made from a live attenuated strain of *M. bovis* named the Bacillus Calmette-Guérin (BCG) after the Frenchmen who developed it. When this is injected into the skin it produces a limited, non-progressive primary infection (a small ulcerating abscess and a reaction in the local lymph node) that heals in a few weeks, leaving behind the protective immunity. It must not be used in patients who are immunodeficient.

Bovine tuberculosis is controlled by the eradication of the disease in cattle. When this is done disease due to *M. bovis* disappears from the human population. Human tuberculosis is controlled by an active programme designed to find cases, with the treatment of any found, to reduce person-to-person spread. The success of this approach will be under threat if multi-resistant bacilli become widespread. Vaccination will then become even more important.

9.2 THE OPPORTUNISTIC MYCOBACTERIA

Minor or subclinical infections with other mycobacteria are common, especially in tropical areas. *M. marinum* is a pathogen of fish that can cause 'swimming pool granulomas'. These are local chronic ulcerative lesions that may affect people who swim in water that contains the organism, or who care for fish in aquariums or fish farms. Infections with opportunistic mycobacteria lead to the development of hypersensitivity to their protein antigens. As the proteins are similar to those of classical tubercle bacilli this can confuse the interpretation of tuberculin tests.

In debilitated or immunocompromised people (for example in AIDS) the opportunistic mycobacteria can cause more severe generalized infections involving the lungs and lymph nodes. Opportunistic mycobacteria are often resistant to a range of antimicrobial drugs, including those normally used for the treatment of tuberculosis.

9.2.1 *M. leprae*

Like tuberculosis, leprosy has a long history (Box 9.4), though because *M. leprae* has not been grown in culture less is known about it. In the tissues *M. leprae* appears as a typical acid-fast bacillus that is located principally in nerves, skin and mucous membranes. The number of them found in these sites reflects the ability of the host to react to the invasion, and this determines the type of disease that is produced.

Box 9.4 Hansen's disease

The first records of leprosy appear in Indian and Chinese writings of about 600 BC. Leprosy was common throughout Europe in the Middle Ages, but by the beginning of the twentieth century it was restricted to parts of Norway and Sweden. *M. leprae* was first seen in 1873 in Norway by Gerald Hansen.

It was known that the disease was contagious and in Europe lepers were outcasts of society. In some places they were made to carry a clapper or bell to give warning of their approach and they were excluded from churches some of which provided narrow slits or 'leper's squints' so they could watch services from the outside. The word leper was also a term of abuse and to avoid the stigma it was at one time proposed that the infection be called 'Hansen's disease'.

WHO estimates that there are 10 million sufferers from leprosy worldwide. In the worst affected places it seems that most of the population has met and acquired hypersensitivity to *M. leprae* by the age of 10, and that 2% of them develop disease as a result.

(a) Pathogenesis

M. leprae gains access to the tissues of a new host through the skin or the mucous membranes, usually of the upper respiratory tract. In most cases a subclinical infection follows with the development of cellular immunity, as in tuberculosis. In a few individuals the bacilli spread and multiply in the sheaths of nerves where damage causes small patches of numbness that may be the first symptom of leprosy.

In those who react more vigorously to the invasion, collections of phagocytic cells develop to enclose the invading mycobacteria. Some of these coalesce to form giant cells, and lymphocytes and macrophages invade the area. A fibrous shell develops round this reaction, to form a tubercle. These cause damage to nerves and lead to widespread loss of sensation and paralysis of muscles. Joints no longer supported by the muscles and tendons that control them suffer progressive damage and extremities not protected from injury by pain are eventually destroyed. Gross deformities are the result. This condition is called **tuberculoid leprosy**.

In those who do not react to the invasion in this way the bacilli multiply much more actively. They soon spill out of nerves into the surrounding tissues, often into the skin. Bacilli collect in enormous numbers (millions in each gram of tissue) and they spread widely. The skin becomes thickened and disfigured. Bacilli may also invade the liver, kidneys and other viscera. This is **lepromatous leprosy**.

Intermediate leprosy consists of a complete spectrum of disease that represents all the possible variations between the tuberculoid and lepromatous extremes just described. The various forms reflect differences in the vigour of patients' responses to the bacillus. To complicate this, patients may change the way they react as their illnesses progress. Symptoms and signs of more extreme lepromatous or tuberculoid disease are then superimposed on each other.

(b) Diagnosis

Swabs or scrapings of the nasal mucosa stained by the acid-fast method will often reveal the presence of AFBs in early lepromatous or intermediate forms of the disease. In lepromatous disease the microscopical examination of a small fragment (biopsy) of skin will show that AFBs are present. In tuberculoid leprosy nerves may be seen to be invaded by AFBs, but the bacilli are usually so scanty that the diagnosis has to rest on the characteristic pattern of the inflammatory reaction. As *M. leprae* cannot be cultured, the demonstration of antimicrobial sensitivity or resistance depends either on difficult animal experiments, or on the taking of serial biopsies from patients. These are examined to measure progress. The degree of fragmentation of bacilli and their eventual disappearance from the tissues is noted when treatment is successful, or a lack of these changes is observed when the bacilli are resistant to the drug that has been used.

(c) Treatment

Dapsone (to which resistance has begun to appear) and rifampicin (which is expensive) are the most important drugs used to treat leprosy. Clofazimine, ethionamide and prothionamide are alternatives. Combinations of these are often employed as in tuberculosis, for the same reason. Treatment must be continued for years, perhaps for 10 or 15 in the case of lepromatous disease.

(d) Epidemiology and control

Transmission is generally agreed to be associated with intimate contact between, in particular, a lepromatous patient and a child, so is common within families. Nasal secretions are one of the most infectious materials. Because *M. leprae* multiplies so slowly, the incubation period is measured in years. Control is attempted by identifying cases of leprosy and treating them to render them non-infectious, particularly when they are of the much more infectious lepromatous type. Prophylaxis with dapsone is sometimes used to protect household contacts of a patient with an infection until the latter has been treated for long enough to become non-infectious. There is evidence that BCG vaccination gives some protection against leprosy.

FURTHER READING

See Appendix B.

10

The enterobacteria
(Salmonella, Shigella, Escherichia, Klebsiella, Proteus, Cibrobacter, Enterobacter, Serratia)

10.1 INTRODUCTION

The enterobacteria are a large group of Gram-negative aerobic and facultatively anaerobic non-sporing bacilli commonly found in human and animal intestines. A few of them are human pathogens and the others are commensals and potential pathogens. Their structures and most of the pathology they cause are similar, so they are conveniently described together. The genera listed at the head of this chapter are those more often found in human infections.

   The outer membranes of the cell walls of Gram-negative bacteria contain their endotoxic component (Chapter 3). Inside these outer membranes are standard bacterial cell walls made of peptidoglycan. Many of the enterobacteria also possess flagella so they can swim about, and some have capsules that may contribute to their pathogenicity.

10.2 THE SALMONELLAS

Salmonellas are motile, Gram-negative rods that share the features just noted. Some authorities recognize over 1700 different named species, while others consider that they are all variants of a single species. The salmonellas are distinguished from other enterobacteria by simple laboratory tests and the many species (or variants) are separated by the antigens they carry on their cell walls and their flagella (Box 10.1). They are all pathogens or potential pathogens. They cause two kinds of disease in humans, with occasional intermediate forms. These are summarized in Box 10.2.

10.2.1 Enteric fever

Individual cases of enteric fever caused by *Salmonella typhi* or by *S. paratyphi* A, B or C are indistinguishable at the bedside, but on average typhoid is

---

Box 10.1 Bacterial antigens

The antigens of Gram-negative bacteria are divided into three categories by their anatomical locations.

1. 'O' antigens are situated in the outer membrane of the bacterial cell walls.
2. 'H' antigens are associated with the flagella, so are only found in bacteria that are motile.
3. 'K' antigens are molecules that make up the capsules of bacteria that possess these structures.

Box 10.2 Human diseases caused by salmonellas

The enteric fevers are caused by strictly human pathogens, *Salmonella typhi* (typhoid fever) and *S. paratyphi* A, B or C (paratyphoid fevers). Disease results when an infectious dose of one of these salmonellas is consumed in food or drink contaminated from a human source. They present as severe septicaemias.

Salmonella food poisoning follows the ingestion of food that contains large numbers of the kinds of salmonellas that come from animals. Most human infections are due to carelessness that converts a small number of salmonellas in food into a toxic dose of them. The infectious process is normally restricted to the gut.

The enteric salmonellas cause severe general disease in normal people so are highly pathogenic. The less pathogenic animal salmonellas only cause local disease in normal people but in those predisposed to infection by debility or immunodeficiency they may spread to cause an enteric-like illness.

---

a more severe infection. The diseases arise when food or drink is consumed that has been contaminated with excreta from a human case of one of the infections, or from a long-term carrier of one of the bacteria. The excreta (nearly always faeces but occasionally urine) may have been diluted in untreated sewage that leaks into a well, or that is discharged into a river or estuary. If a few of the bacteria reach food they can multiply there to develop into an infectious dose. Paratyphoid A is more common in the Indian subcontinent and SE Asia, while paratyphoid B and typhoid itself are found worldwide. Paratyphoid C is rare.

Before drugs were available to treat these infections they ran a course that was divided into the five stages outlined in Box 10.3. The later stages are modified when infections are treated with an antimicrobial to which the salmonella is sensitive.

(a) Diagnosis, treatment and control

The salmonella responsible may be isolated by culture from the blood, faeces or urine. If they are counted, the white cells in the blood are found to be

---

Box 10.3 The enteric fevers

Before antimicrobial treatment was available, or now in the absence of it, these illnesses pass through five stages.

1. **Incubation**. Salmonellas in food or drink that survive the gastric acid invade the lymphatic tissues of the jejunum, ileum and colon, and multiply there. This stage lasts between one and three weeks: 10–12 days is usual.

2. **Invasion**. The patient is ill and feverish, with 'rose spots' in the skin. Constipation is common. This stage lasts for about a week and the fever rises day by day to produce a temperature chart that resembles a staircase.

3. **Fastigium**. The temperature reaches a high plateau and the patient is toxic and drowsy. This stage also lasts about a week.

4. **Recovery or complications**. The defences begin to overcome the invader, but the patient is severely weakened. Ulcers in the gut may perforate to cause peritonitis, or they may bleed. The patient is very ill and in no state to withstand either event. Before antimicrobials appeared this was the time when skilled nursing care might decide if a patient lived or died.

5. **Convalescence**. A slow improvement follows, with possible relapses. The salmonellas are gradually eliminated but still appear in the faeces for a time. This convalescent carriage may become chronic if a salmonella settles in an abnormal gallbladder or urinary tract. When this happens they are shed in the faeces or urine, probably for years.

---

reduced in number, an unusual finding in a bacterial infection. Chloramphenicol, cotrimoxazole, ampicillin and more recently ciprofloxacin have been used successfully in the treatment of the enteric fevers. Plasmid-mediated resistance to chloramphenicol has appeared in some parts of the world, sometimes with resistance to ampicillin and cotrimoxazole as well. Although laboratory tests show that salmonellas are sensitive to the aminoglycosides, they and the cephalosporins have proved disappointing in use, probably because they fail to reach bacteria that multiply inside cells.

Box 10.4 Royal typhoid

On 14 December 1861, at the age of 42, Queen Victoria's consort Prince Albert died of enteric fever at Windsor Castle. Ten years later their son Edward, Prince of Wales and heir to the British throne, nearly died of the same disease. At the time the arrangements for the disposal of sewage at Windsor Castle were much as they had been in medieval times.

Estimates of the incidence of typhoid fever in different parts of the world vary between one per million and six per thousand of the population each year. This variation reflects differences in socio-economic development and hygiene in the disposal of excreta and the provision of safe food and drink (Box 10.4).

The effectiveness of the vaccines available varies with the number of *S. typhi* consumed. A vaccinated individual may not be able to resist a large infectious dose.

Patients with enteric fever (or carriers of it) can only transmit the infection to others through their excreta. Simple hygienic disposal of these can break the chain of infection. Elaborate isolation of patients is not necessary provided their excreta are handled with reasonable care. Chronic carriers should not be allowed to prepare food intended for others. If all these individuals could be detected and freed of their carriage the enteric fevers would disappear.

10.2.2 Salmonella food poisoning

Food poisoning may otherwise be known as gastroenteritis, enteritis or enterocolitis, depending on the part of the gut affected. The variety due to the salmonellas is an acute infection with an incubation period of 12–48 hours. The disease has a sudden onset with diarrhoea and abdominal pain, often with nausea and sometimes, vomiting. Dehydration due to the diarrhoea may be severe, especially in infants. Fever is common, and anorexia and loose stools may persist for several days. In debilitated patients the typical enterocolitis may develop into a more severe illness with a septicaemia similar to enteric fever. Unusually, salmonellas may localize in other parts of the body to cause abscesses, arthritis, cholecystitis, endocarditis, meningitis, pyelonenephritis or osteomyelitis.

In uncomplicated food poisoning the causative salmonella is excreted in large numbers in the faeces for the first few days of the illness. The number declines as time passes, and excretion usually ceases after one or two weeks. Long-term chronic carriage of non-enteric salmonellas is rare in humans, though it is common in animals, particularly if these are infected at or about birth.

(a) Diagnosis

Salmonellas may be isolated from the faeces in cases of food poisoning, and from the food that has caused it. They may be isolated from other sites when the infection has spread outside the gut. In the laboratory selective methods of culture are used to separate salmonellas from the confusing array of similar bacteria found in the faeces.

(b) Treatment

Reassurance, and if the diarrhoea is severe, rehydration are all that is necessary in most cases. Antimicrobials do not seem to shorten the acute stage of uncomplicated food poisoning, and the unwise use of them commonly lengthens the period during which salmonellas are excreted in the faeces after recovery. A salmonella septicaemia is dealt with in the same way as enteric fever, and local salmonella sepsis is treated according to the site involved.

(c) Epidemiology and control

Salmonella enterocolitis is an important disease worldwide, as indicated by the range of the geographical names that have been applied to their species, or serotype varieties. A selection from only the first three letters of the alphabet includes Adelaide, Amsterdam, Atlanta, Belfast, Berlin, Brazil, Brunei, California, Colombo, and Detroit.

The disease results from the ingestion of food or drink each gram of which contains 100 000 or more salmonellas. Meat and poultry may be contaminated because they came from infected or carrier animals. Mild or symptomless infections and the carrier state are common among commercial herds of cattle, sheep or pigs, or in flocks of chickens, ducks, geese or turkeys. Uncontaminated food that comes into direct or indirect contact with a contaminated product is contamined in turn (Chapter 28). Cross-contamination can happen in the abattoir or slaughterhouse, or in a warehouse, shop, restaurant or in the home, perhaps in refrigerators. Milk and eggs may be contaminated at source, and shellfish harvested from sewage-polluted water may contain an infective dose of salmonellas. The recycling of waste food and parts of animal carcasses in commercially produced animal feed ensures that infection remains endemic.

The contamination of food by human carriers is possible, as happens in enteric fever, but investigation of outbreaks of salmonella food poisoning nearly always shows that the infection originated in an animal or avian source. Improper storage of food allows small non-infective populations of salmonellas to multiply to reach toxic levels. Inadequate or non-existent refrigeration is the usual cause of this. In catering establishments cooked and uncooked foods should not share the same refrigerator. The number of salmonella present in an item of food, the amount of it, and the number of people who eat it are the factors that determine if food poisoning afflicts a single individual, or several hundred people.

Food handlers usually sample the food they prepare so are likely to suffer from salmonella infections at the same time as others who eat the contaminated food. As a consequence they are sometimes blamed as sources of outbreaks of food poisoning when in fact they are just other victims of it. Even so it is wise to exclude food-handlers with food poisoning from work until they have recovered, and cultures of their stools are negative for salmonellas. The same applies to health care workers, particularly if they are employed in intensive care or paediatric areas. In other cases responsible people with hygienic habits may return to work when their diarrhoea has ceased, provided they do not handle food for others or care for debilitated people or groups of very young children. In these cases 'clearance' stool cultures are not necessary.

In hospitals salmonella infections may spread from patient to patient by direct contact, that is, by routes that do not involve food. Patients whose production of gastric acid has been reduced because of illness or by therapy may be infected by small, easily transmitted numbers of salmonellas of no consequence in other circumstances. Once infected such people can be the focus of a hospital outbreak that is difficult to control.

10.3 THE SHIGELLAS

Shigellosis or bacillary dysentery is caused by bacteria of the genus *Shigella*. They have no flagella so are non-motile, and they are distinguished from other enterobacteria by their metabolic activities and antigenic structures.

These microbes live in the large intestine of humans and other primates. They rarely spread beyond the lining of the colon, and are shed in the faeces. They are highly infectious by the faecal-oral route as the minimum infective dose (perhaps 1000 shigellas) is small compared with the salmonellas. After an incubation period of one to three days they cause an acute colitis, of variable severity. The patient usually has a temperature, and may also feel nauseated and vomit. A very large number of stools may be passed each day. Shigellas produce pathology by local action with irritation and damage to the wall of the colon. Shiga's bacillus (*Shigella dysenteriae* type 1, Box 10.5) also produces an exotoxin. This shigella is the cause of the most severe form of dysentery, with a significant mortality. The least pathogenic, *Sh. sonnei*, usually produces a very mild disease.

Box 10.5 The bloody flux

A shigella was first isolated by Kiyoshi Shiga in 1898. At the time there was an epidemic of dysentery in Japan in which about 100 000 people were attacked, and 25 000 died.

Although now less common in developed countries dysentery is still a major cause of illness and death in poorer parts of the world. It may affect adults when hygiene breaks down as the result of a disaster, particularly in crowded makeshift encampments, but it is primarily a disease of children.

Shigellosis is unusual in breast-fed infants, but it becomes more common after weaning and in some parts of the world each one-to three-year-old may suffer more than one attack of it every year. Among the malnourished the disease can be fatal. The old idea that the bowel should be rested in dysentery is bad advice. As well as fluid replacement some protein and calorie intake is necessary. The best antimicrobial is ampicillin, if the shigella is sensitive, but resistance to this drug and the more usual alternatives has become common in some parts of the world.

(a) Diagnosis and treatment

The stool in an established case of shigellosis is characteristic, though not diagnostic, as similar stools are produced in ulcerative colitis. Early on it is watery (profusely so in *Sh. dysenteriae* type 1 infections) with flakes of mucus. Later it becomes small in volume and consists largely of blood-stained mucus that sticks to the bedpan, typically without faecal odour or colour. In very severe cases it may resemble pus, and contain necrotic mucosa.

Culture of the stool on appropriate selective media will reveal the causative *Shigella*, while microscopy shows the presence of large numbers of pus cells, a feature that helps to distinguish it from amoebic dysentery. Fluid replacement is an important part of treatment and antimicrobials are used to limit damage to the colonic mucosa in severe infections, though as with salmonella enterocolitis they do not seem to make any difference in mild cases. Some of the more pathogenic shigellas may be multiply resistant to antimicrobials.

(b) Epidemiology and control

Shigellosis is a human disease, with no significant animal reservoir. The disease persists as a result of chronic carriage in a proportion of patients, who may continue to excrete the organism for a year or more after recovery. In developed societies the infection is often due to *Sh. sonnei*. This is typically a disease of young children, particularly at the nursery or primary school stage, when toilet training is incomplete. Transmission is by unwashed hands and in food; it is also traditional to blame flies. The more severe forms are seen in less developed countries (Box 10.5). Major epidemics still occur. In 1969 in Central America there were 100 000 cases with at least 8000 deaths, and in 1984 in West Bengal Shiga's bacillus killed over 2000 people in a few months.

10.4 THE OTHER ENTEROBACTERIA

The remaining enterobacteria comprise a collection of similar microbes with properties that vary only a little. Some may act as pathogens in the gastrointestinal canal, but most are commensals that only cause infections when they spread outside the gut and invade areas of the body from which they are normally excluded.

10.4.1 Eschierichia coli

E. coli is a normal, universal inhabitant of the human intestine. Here it may cause gastroenteritis, or it can act as a pathogen if it spreads outside it. Diarrhoea may result when a newly-acquired strain of *E. coli* invades the intestinal mucous membrane or produces an enterotoxin. Strains of *E. coli* with these properties are common causes of diarrhoea in infants, and of 'travellers' diarrhoea' ('Delhi belly', 'Montezuma's revenge', 'Pharaoh's curse', 'tropical trots', etc.). One particular type of *E. coli* is a cause of the haemolytic-uraemic syndrome, in which gastroenteritis is accompanied by severe anaemia due to the destruction of red blood cells, and renal failure.

When some event opens a pathway into a normally bacteria-free area, the ubiquitous *E. coli* is always at hand to take advantage of it. This is why *E. coli* is the most common cause of urinary tract infection (UTI). UTIs are common among young women, especially in pregnancy. They are also seen in the very young (perhaps associated with abnormalities of the urinary tract) and in the elderly, in whom it is a regular complication of various gynaecological problems and incontinence in females, and hypertrophy of the prostate in males. Prolonged urinary catheterization is always accompanied by an infection, the initial cause of which is often *E. coli*.

E. coli may also cause wound infections. This happens particularly after operations on the bowel and when anaerobic organisms are often present at the same time. In neonates it can cause meningitis, and it is commonly found in the septicaemias that complicate local *E. coli* infections, often of the urinary tract. A man with prostatic hypertrophy and UTI due to *E. coli* may develop a rapidly fatal septicaemia after catheterization or cystoscopy. An old, incontinent woman who is catheterized for convenience may suffer the same fate.

(a) Diagnosis

Standard cultural methods allow the isolation, identification and sensitivity testing of *E. coli*. It is sometimes difficult to assess the significance of this organism when it is found in a clinical specimen. This is because colonizations of sites other than the intestine are common, particularly where the normal flora has been disturbed by antimicrobial therapy.

(b) Treatment

In cases of *E. coli* diarrhoea, fluid replacement is usually all that is needed. The troublesome travellers' diarrhoea has been treated with many different preparations, including antimicrobial drugs. Manufacturers may make extravagant claims for their products but they seem to be of little use in practice. Wound or other infections are treated as indicated by the circumstances. Many strains of *E. coli* have become resistant to antimicrobials, reflecting the wide use of these drugs and the oral route of administraton in animals as well as humans.

10.5 THE REMAINING ENTEROBACTERIA

The rest of the enterobacteria listed at the beginning of this chapter can also cause infections of the urinary tract or tissues. All of them have a well-marked ability to develop resistance to antimicrobial drugs. In hospitals it is common to find patients whose gastrointestinal flora has changed, in whom the usually dominant and more sensitive *E. coli* has been displaced by another of the enterobacteria, perhaps a more resistant klebsiella. The new inhabitant may then spread to cause infections in the manner described for *E. coli*, though they are now more difficult to treat. These microbes, sometimes called 'hospital strains' of bacteria, are often found as the causes of hospital-acquired infections.

FURTHER READING

See Appendix B.

The curved rods 11
(Vibrio, Campylobacter, Helicobacter, Aeromonas, Plesiomonas)

11.1 THE VIBRIOS

Opinions differ about which organisms to include with the curved rods as 'vibrios', and the word 'cholera' is not interpreted straightforwardly. The vibrios may be defined loosely (and conveniently) to include Gram-negative bacilli that are motile by means of flagella attached at one end ('polar flagella'), that often appear curved or comma-shaped, and that share certain other characteristics of a technical nature. The important members of the group formed by this definition include the bacterial genera named above. Their activities are summarized in Box 11.1.

Box 11.1 The 'vibrios'

Vibrio cholerae
Serovar O1, subtype Inaba or Ogawa, biovar classical or eltor, the causes of 'true' cholera, as internationally recognized. In future this group will include serovar O139.
   Non-cholera vibrios (NCVs) serovars O2-O138 may cause sporadic cholera-like infections.

V. parahaemolyticus
A salt-tolerant marine vibrio. It causes food poisoning that is particularly associated with the consumption of shellfish.

Campylobacter, especially *C. jejuni*.
A microaerophilic, carbon dioxide dependent vibrio that is an important cause of acute enteritis.

Helicobacter pylori
A recently-rediscovered bacterium found in the stomach of patients with gastritis and duodenal ulceration. It is generally accepted that it acts as a pathogen at this site.

Aeromonas and *Plesiomonas*
Species of these genera have been isolated from faeces in cases of diarrhoea, and from wounds. Their pathogenic status is debated.

Some vibrios, campylobacters and helicobacters are important human pathogens. Although both *Aeromonas* and *Plesiomonas* are regularly found in cases of enteritis in the absence of other causes not everyone will accept them as pathogens, though the evidence for *Aeromonas* at least is strong.

Cholera is an infection in which dehydration and loss of essential salts (electrolytes) result from diarrhoea caused by toxigenic bacteria. If this broad definition were accepted, then the disease is caused by a number of different microbes, including strains of *Escherichia coli* that produce a cholera-like enterotoxin. There are sound practical and legalistic reasons for the use of a much narrower definition that includes only the epidemic disease caused by those strains of *Vibrio cholerae* that possess a special transmissibility. This has been accepted by international agreement and the World Health Organisation (WHO) so that countries are able to react in a standard way when 'cholera' is reported. The terms 'paracholera' or 'cholera-like disease' are applied to non-epidemic though clinically similar conditions caused by other microbes.

This is not the end of the difficulty. Experts have included organisms that may cause sporadic cholera-like disease within the species *V. cholerae*. The effect of this is that some strains of *V. cholerae* cause 'true' (internationally recognized) cholera, while others cause sporadic and non-epidemic cholera-like diseases that are individually indistinguishable from the 'true' infection.

The species *V. cholerae* is divided into over 100 varieties by antigenic differences among their 'O' antigens (Box 10.1). Each of the varieties is identified when it reacts with serum that contains an antibody prepared against its own unique antigen. Until recently *V. cholerae* 'serovar' O1 included all the microbes that cause epidemic cholera. In 1992 outbreaks of cholera were detected along the coast of the Bay of Bengal in India and Bangladesh that were due to a new *V. cholerae* serovar, numbered O139. This variant is clearly associated with epidemic disease so a new bacterial cause of true cholera has emerged. Other serovars of *V. cholerae* that are either non-pathogenic or cause paracholera are called non-cholera vibrios (NCVs).

V. cholerae serovar O1 includes two sub-species or biotypes (biovars) that are separated by laboratory tests. These are labelled 'classical' and 'eltor'. Either biovar may carry one or other of two minor antigens called 'Inaba' or 'Ogawa', in addition to the O1 antigen. There are thus a total of four kinds of *V. cholerae* O1. The full description of a vibrio isolated from a case of cholera is *V. cholerae* serovar O1 (Inaba or Ogawa), biovar classical or eltor, now to include serovar O139.

11.1.1 Cholera

Cholera is defined officially as an acute diarrhoea caused by *V. cholerae*, serovar O1 or O139. In the centuries before 1817 the disease seems to have been confined to the Bay of Bengal. Since 1817 there have been six pandemics in which cholera has spread round the world from its primary focus on the delta of the Ganga and Brahmaputra rivers. At different times these pandemics have involved the rest of Asia, Africa, the Mediterranean, Europe and the

Americas. The cholera vibrio was first isolated by Robert Koch in Egypt during the fifth pandemic (1883–1896) (Box 11.2). Until 1961 cases of the disease were caused by this original or 'classical' biovar of *V. cholerae* O1. In 1961 a fundamental change took place, and the eltor vibrio emerged as the major pathogen.

Box 11.2 The race for the cholera vibrio

The Franco-Prussian war of 1870–71 left a legacy of hatred among Frenchmen. For Louis Pasteur this added a special extra dimension to the rivalry that existed between himself and the German, Robert Koch. Both men were keen to discover the cause of cholera. An opportunity presented when the disease appeared in Egypt in 1883. Pasteur sent a team under Emile Roux, and Koch headed one of his own. The race was on.

At one time it seemed that the French had won, but Koch realized that they had mistaken blood platelets for the cause of the disease. A little later Koch learned that one of the French team had fallen ill, with cholera. Koch visited the Frenchman and when asked if what they had seen was the cause of the disease, he said yes. The man died soon afterwards. In the end it was Koch who won the race.

In 1905 an atypical vibrio was isolated from individuals making their pilgrimage to Mecca. It was called the 'eltor vibrio' because the isolation was made in the El Tor quarantine station in Sinai. At the time it was a curiosity because although it closely resembled *V. cholerae* it did not seem to cause disease. In the 1930s in the Celebes (now Sulawesi in Indonesia) and later elsewhere, the eltor vibrio was found in cases of non-epidemic cholera-like disease. In 1961 the eltor vibrio acquired the ability to spread in an epidemic fashion. *V. cholerae* O1 biovar eltor rapidly replaced the classical variety, which for a time almost disappeared. The eltor vibrio spread to Hong Kong, the Philippines, Korea, China and through the rest of SE Asia to reach India, the Middle East, the USSR, Africa, Europe and finally the Americas. It became the cause of the current, seventh cholera pandemic. It is possible that *V. cholerae* serovar O139 will be the cause of the eighth (Albert *et al.*, 1993; Bhattacharya *et al.*, 1993; Shimada *et al.*, 1993; Swerdlow & Ries, 1993).

(a) Pathogenesis

One to five days after an infectious dose of *V. cholerae* has been ingested, the individual suffers the acute onset of a profuse, painless, watery diarrhoea, perhaps with vomiting. As much as 15–20 litres of fluid may be lost in this way in 24 hours, and the stool comes to resemble the water in which rice has been cooked ('rice-water stool'). At this rate the daily loss represents nearly half the volume of water present in the body. If it and the electrolytes are not replaced death follows in a few hours (Box 11.3). Dehydration is associated with painful muscle cramps, the eyes are shrunken, the elasticity of the skin is lost, the blood pressure falls, and no urine is passed. Finally the circulation fails and the patient lapses into coma and dies.

Box 11.3 A quick death

Early in 1857, a 'promising young soldier', Corporal Sweeney, was on the march with his regiment towards Delhi, the capital of India. One day at the end of the morning he told his company commander that two of his men had died of cholera. A funeral was arranged for 6.00 p.m. Corporal Sweeney fell ill and died in time to be buried with them.

Dramatic events of this sort form a very small part of an epidemic of cholera. Mild or subclinical infections are much more common. The disease tends to be more severe in individuals predisposed by malnutrition or other illnesses. People with hypo- or achlorhydria are more easily infected because hydrochloric acid in the stomach defends the gut from bacteria invasion. Once past this barrier *V. cholerae* attaches itself to the wall of the small intestine, where it multiplies and producs an enterotoxin. This binds to receptors on, and then enters the cells of the lining of the small intestine. Inside these cells the toxin triggers actions that alter their permeability. Water and electrolytes that are normally retained in the blood and tissue fluids pour into the gut and are lost as diarrhoea.

(b) Diagnosis and treatment

In the laboratory *V. cholerae* is easily isolated from the faeces of infected patients, provided the appropriate selective methods of culture are used. Treatment is by the replacement of the fluid and electrolytes as soon as possible after they have been lost. If dehydration is already severe these may have to be given intravenously, in large volumes. As much as 80 litres of fluid may be required over a period of three or four days. This is equivalent to the contents of 10 ordinary buckets. In most cases intravenous therapy can be supplemented, or replaced, by oral treatment. This is just as well as the intravenous fluid requirements in an epidemic area are beyond the capacity of many medical supply organizations.

For oral treatment a balanced solution of salts is used, with glucose, sugar or starch added to encourage their absorption (Box 11.4). Oral rehydration solution (ORS) is a good, safe alternative to parenteral therapy. Where medical help or intravenous fluids are unavailable ORS can be life-saving. The solution is easy to prepare and the constituents cheap and readily stored. The same mixture may be used to correct the dehydration that results from any infective diarrhoea. Generous rehydration should begin as soon as possible. The volume required can be estimated by feeling the patient's pulse and watching the condition of the skin, and by matching the volume of fluid lost as diarrhoea, with a little over. If swallowing is a problem it may be given by naso-gastric tube.

Box 11.4 Oral rehydration solution

The 1985 WHO oral rehydration solution (ORS) contains, in grams per litre of boiled water, sodium chloride 3.5, trisodium citrate (dihydrate) 2.9, potassium chloride 1.5, and (anhydrous) glucose, 20.0. The glucose may be replaced by 40 g/l of sugar of 50 g/l of powdered rice, boiled in the water. Sodium bicarbonate 2.5 g/l may be used in place of trisodium citrate.

The antimicrobial tetracycline can be given at the same time to slow down the multiplication of vibrios in the gut. This limits the diarrhoea and so reduces the volume of fluid that has to be replaced.

(c) Epidemiology and control

The present seventh pandemic has reawakened interest in the epidemiology of cholera. At the time of the third pandemic in 1855 John Snow removed the handle of a pump in Broad Street, London, and apparently halted an outbreak of cholera. He did this as the result of a study that is now regarded as an epidemiological classic. The idea that cholera is spread by water took firm root. Although this is true it is not the whole truth. The number of cholera vibrios found in a drinkable volume of water, even at the height of an epidemic, does not approach an infectious dose for a person with normal gastric acidity. Vibrios do not multiply in water, but they can do so in food. If unrefrigerated food is contaminated by vibrios from the water used to prepare it, an infectious dose is readily achieved in a short time. This food-borne route of infection plays an important part in the epidemiology of cholera.

Most patients with cholera stop excreting vibrios after a few days. A few individuals continue to excrete them for some months. A chronic infection of the biliary system with intermittent shedding of vibrios into the faeces for many years has been described, but it is rare.

Cholera may be controlled by breaking the faecal-oral pathway of infection. Effective sewage disposal, clean water and the hygienic handling of food are the keys. An infection with *V. cholerae* protects against reinfection for only a few months though it may still lessen the severity of a subsequent attack. The vaccines generally available are somewhat less effective than this, so they are not recommended for mass use in the control of epidemics. They may give partial, short-term protection to non-immune individuals visiting highly endemic areas. New experimental vaccines containing the non-toxic part of cholera toxin show more promise (Box 3.3). Tetracycline has been used prophylactically in household contacts of cases of cholera.

11.1.2 *Vibrio parahaemolyticus*

V. parahaemolyticus was first isolated in Japan in the early 1960s. Since then it has been recognized as a common cause of food poisoning. The vibrio is found in a variety of seafoods, and outbreaks of disease due to it have been reported from all parts of the world.

V. parahaemolyticus causes a watery diarrhoea accompanied by cramping abdominal pain perhaps with nausea, vomiting, fever and headache, lasting for a few days. The infection is rarely severe, but the organism can cause a septicaemia. The incubation period is usually between 12 and 24 hours, and the illness follows the consumption of raw or lightly cooked fish or shellfish. Shellfish exposed to boiling water inside their shells for only two or three minutes may still contain large numbers of living vibrios.

The diagnosis of this form of food poisoning is by culture of the faeces, using special media. As befits a marine species *V. parahaemolyticus* is unable to grow on ordinary laboratory media because they contain insufficient salt. This property helps to distinguish this vibrio from *V. cholerae*, and it also ensured that it remained undiscovered until quite recently.

11.2 *CAMPYLOBACTER JEJUNI*

Campylobacters were originally thought of as animal pathogens. More recently some of them, and in particular *C. jejuni*, have been recognized in most parts of the world as very common causes of human enteritis. The organism is acquired from animal sources through food or drink. It multiplies in the small intestine to produce an inflammatory reaction, and diarrhoea. Occasionally it may spread into the blood stream to cause a more serious enteric fever-like disease.

Diagnosis is by culture of the faeces using special media incubated in an atmosphere with reduced oxygen and extra carbon dioxide. The importance of campylobacters as human pathogens was missed for a long time because these conditions were not provided in routine laboratory practice. Erythromycin is an effective treatment for a severe infection.

11.3 *HELICOBACTER PYLORI*

In 1983 a curved or spiral Gram-negative bacterium first noticed many years before was rediscovered in the gastric mucosa of patients with gastritis. Eventually it was grown in the laboratory and is now named *H. pylori*. The strong association with gastritis led to the theory that this helicobacter plays a part in the aetiology of gastritis and duodenal ulceration. This is now generally accepted.

FURTHER READING

1. Albert, M.J., Siddique, A.K., Islam, M.S. *et al.* (1993) Large outbreak of clinical cholera due to *Vibrio cholerae* non-O1 in Bangladesh. *Lancet*, **341**, 704.
2. Bhattacharya, M.K., Bhattacharya, S.K., Garg, S. *et al.* (1993) Outbreak of *Vibrio cholerae* non-O1 in India and Bangladesh. *Lancet*, **341**, 1346–7.
3. Shimada, T., Balakrish Nair, G., Deb, B.C. *et al.* (1993) Outbreak of *Vibrio cholerae* non-O1 in India and Bangladesh. *Lancet*, **341**, 1347.
4. Swerdlow, D.L. and Ries, A.A. (1993) *Vibrio cholerae* non-O1 – the eighth pandemic? *Lancet*, **342**, 382–3.
5. See Appendix B.

12 The pseudomonads and legionellas
(Pseudomonas, Legionella)

Bacteria of these genera are non-sporing, Gram-negative and rod-shaped. They are widely distributed in the wet parts of the environment. Some of them are potential human pathogens.

12.1 THE PSEUDOMONADS

Pseudomonads are distinguished from other Gram-negative rods by being obligate aerobes, by motility with polar flagella, and by the possession of certain metabolic features of interest to microbiologists. Several species make bright coloured pigments and produce extracellular antimicrobial substances that inhibit the growth of other bacteria. To survive they must themselves resist these substances, so they are also resistant to many of the antimicrobial drugs in common use. In addition they can withstand the action of some disinfectants, though they are easily killed by heat. Infections, when they arise, are difficult to treat, and some pseudomonads can grow in the weaker solutions of disinfectants that are used on the skin and in liquid medicaments such as eye drops. The species of importance in medical microbiology are *Pseudomonas aeruginosa*, *Ps. mallei* and *Ps. pseudomallei*. *Ps. mallei* is the cause of an infection called 'glanders' in horses, mules and donkeys. This sometimes spreads to the humans who work with them, though it is now an exceedingly rare disease. Pseudomonads of species other than those listed are sometimes implicated in infections, particularly if they are acquired in hospitals.

12.1.1 *Pseudomonas aeruginosa*

(a) Pathogenesis

This microbe usually produces pigments, most notably a strong green one. This is the cause of the blue-green pus that may be produced in infections due to it. *Ps. aeruginosa* is sometimes found in small numbers as part of the normal flora of the gut (Box 12.1). Otherwise it colonizes or infects

Box 12.1 *Pseudomonas aeruginosa*

This is a heavily armed bacterium. It is armed with pili for attachment, a flagellum for motility, and a heavy mucoid capsule inside which it can hide from body defences and antimicrobials. It is naturally resistant to many disinfectants and antibacterial drugs, and it makes powerful antimicrobial substances of its own. One of these resembles diphtheria toxin but fortunately this is ill-equipped to attack human cells.

In practice *Ps. aeruginosa* only attacks tissues that are already compromised in some way, so it causes infections in hospitals twice as often as in the community. The carriage of *Ps. aeruginosa* in the intestine may be about 20% in the community, 30%–40% among ordinary patients in hospital, rising to 90% after seven days in an intensive care unit.

areas of the body where damage or disease has produced conditions favourable for it. Such conditions exist when a surface that is normally dry becomes wet. This happens when surfaces are denuded of their surface epithelium as in burns, ulcers, pressure sores and in the external auditory meatus in chronic otitis externa (Box 12.2). Other favourable sites develop when mucous surfaces have been stripped of their ciliated epithelium or are otherwise deprived of the mechanisms by which they normally keep themselves free of microbes. This is seen in the lungs in cystic fibrosis and bronchiectasis, in the middle ear in chronic otitis media, and in the urinary tract when it is obstructed or catheterized for a long time. *Ps. aeruginosa* may also infect wounds and the eyes, and cause diarrhoea in infants.

Box 12.2 Swimmers' ear

Chronic infection of the external auditory meatus ('otitis externa') is a problem for some swimmers. Water in a swimming pool rapidly develops a thin scum on its surface. This is a layer of oil and grease, plus skin squames and other debris. The fats are derived from normal skin and from cosmetic and other materials applied to it. Bacteria congregate in this layer. If *Ps. aeruginosa* is present they are not only naturally somewhat resistant to the chlorine commonly used as a disinfectant but are further protected from it by the grease. Swimmers' heads are at just the right level for this surface layer to flow into their ears. *Ps. aeruginosa* is the most common cause of 'swimmers' ear'.

In any of these situations it may be found that a pre-existing normal colonization or infection with a bacterium that is sensitive to antimicrobials has been disturbed by antimicrobial treatment. This creates a space for the much more resistant *Ps. aeruginosa*. The ease with which *Ps. aeruginosa* can grow in water or dilute solutions of disinfectants is important in hospitals, medical centres and clinics. Contaminated water may collect in respirators or humidifiers in intensive care units, and this may be inhaled. Intravascular lines used for hydration, nutrition or monitoring may become contaminated with *Ps. aeruginosa*, which is then provided with a highway directly into the cardiovascular system. A catheter in the urinary tract can act in the same way. Eye drops or dilute solutions of disinfectants can become cultures of a pseudomonas. Tanks of water used to rinse endoscopes after they have been disinfected may be colonized by pseudomonads, to recontaminate the apparatus before it is used.

Ps. aeruginosa is a much more invasive pathogen in debilitated or immunocompromised individuals. Patients with leukaemia whose white blood cells have been destroyed as a part of their treatment are at particular risk. Any local colonization or infection with *Ps. aeruginosa* may then be the source of a potentially lethal septicaemia.

(b) Diagnosis, treatment and control

The organism can be isolated and identified by normal bacteriological methods from specimens collected according to the clinical indications. Once isolated antimicrobial sensitivity tests provide a guide to the choice of therapy, particularly in hospitals where resistant strains may be found. Because these infections can be severe, combinations of an aminoglycoside with an antipseudomonal penicillin like piperacillin may be used for treatment. Some of the later cephalosporins (ceftazidime, for example) and the quinolones are also effective.

Because *Ps. aeruginosa* is quite commonly found in the gut, and because it can survive and even multiply in the moist environment, it is a difficult organism to control. Anything that can be should be kept dry, and wet areas, particularly if they are to come into contact with patients, require special attention. Care is required when using liquid medicaments, particularly those that contain 'preservatives'.

12.1.2 *Ps. pseudomallei*

(a) Pathogenesis

This bacterium is found widely distributed in soil and water in the humid tropical areas of SE Asia and in some other places. It is the cause of what may be thought of as a rare life-threatening human disease, melioidosis. In fact the small number of cases of melioidosis that are diagnosed represent the tip of an iceberg. The submerged part is made up of a much larger number of mild or subclinical infections. In some areas 5–20% of agricultural workers have antibody to *Ps. pseudomallei* though without a history that suggests that they have ever suffered from melioidosis. Debility, old age and immuno-compromise predispose to the more serious variety of the infection. The disease is similar to human glanders, presenting either as a rapidly fatal septicaemia, or it may run a more chronic course resembling tuberculosis, with pneumonia, multiple abcesses and osteomyelitis. The organism can remain latent in the body for a long time after initial exposure to it. In consequence the incuba-tion period may run into years, until time or chance lowers the resistance of the individual so that the weakly pathogenic bacterium can emerge and begin to multiply. The disease may also affect wild and domestic animals (Box 12.3).

Box 12.3 Melioidosis

All four gorillas imported into Singapore for their zoo died of melioidosis within a few months of arrival, yet this severe disease is rare among the human population of the island. The bacterial cause, *Ps. pseudomallei*, is widely distributed in SE Asia, and asymptomatic infections due to it are common. It has been estimated that over 200 000 of the 2.5 million Americans who served in the US forces in Vietnam were infected, though only 343 cases and 36 deaths were reported at the time. The fact that 200 000 people may be carriers of the organism, some of whom may suffer from it later, has been called 'the Vietnam time-bomb'.

(b) Diagnosis, treatment and control

The diagnosis is made by isolating *Ps. pseudomallei* from blood cultures or other appropriate specimens, or by the identification of antibodies to the organism in the patient's serum. Treatment is with ceftazidime initially, followed by other antimicrobial drugs given for an extended period. The infection is environmental and it does not commonly pass from person to person, so control is impractical. Like *Ps. aeruginosa*, *Ps. pseudomallei* can multiply in dilute solutions of disinfectants, and these have caused hospital outbreaks of the disease.

12.2 THE LEGIONELLAS

The legionellas are rod-shaped organisms that show up poorly in clinical material when these are stained with some red dyes though with others they are clearly Gram-negative. They grow slowly in the laboratory on special media, taking between two and five days to form visible colonies. The genus includes a growing list of related species, of which the most important is *L. pneumophila* (Box 12.4). *L. pneumophila* is itself divided into a number of serogroups according to the possession of different antigens. Members of serogroup 1 appear to be the most pathogenic.

Legionellas are difficult to grow in the laboratory but in the environment they multiply actively in naturally dirty water at temperatures between 25°C and 45°C. They may be found in lakes and streams, but the conditions in artificial systems, in particular air-conditioning plants and the complex plumbing of large modern buildings, appear to suit them best. The reason for the paradox is unclear, but in order to grow, the legionellas seem to depend on the presence of other microbes, particularly amoebae, in the same water. The fact that legionellas can multiply inside amoebae coincides

---

Box 12.4 Legionnaire's disease

In July and August 1976, the American Legion held a convention in Philadelphia, USA. This was accompanied by an outbreak of over 200 cases of severe respiratory illness, with 34 deaths. Most of those affected were ex-servicemen who had attended the convention. This is why the disease became known as Legionnaire's disease, though the name legionellosis is now preferred. Epidemiological studies suggested that the disease had been acquired by exposure in a hotel lobby in the city. Among those who had been in the area, older people seemed more likely to be ill.

The initial search for the cause was negative. Finally in 1977 a bacterium was found in guinea pigs inoculated with post-mortem material taken from a case. Eventually a culture medium was devised that could support the growth of what was eventually called *Legionella pneumophila*. It proved to be the first member of a new bacterial genus that had eluded discovery because of unusual nutritional requirements, and because the disease it caused had not previously attracted attention (see Fraser, *et al.*, 1977).

---

with the observation that in the infections they cause they live and multiply inside human phagocytic cells.

(a) Pathogenesis

Although legionellas are regularly found in the immediate environment of man, significant outbreaks of infection due to them are rare. The factor or factors that determine virulence are not known. What is clear is that *L. pneumophila* of group 1 are of low, and that other legionellas are of even lower, virulence. Exposures that do not cause infection, or that are followed by subclinical or mild infections, must be common. 'Pontiac fever' is a mild influenza-like illness, not accompanied by pneumonia, that is due to heavy exposure to legionellas. More severe legionellosis may develop in individuals predisposed by smoking, chronic lung disease, alcoholism, old age or immunosuppression. Fully developed legionellosis is marked by progression from an initial influenzal phase to severe pneumonia of a patchy, atypical kind, associated with a polymorphonuclear leucocytosis, altered liver function, mental disturbance, perhaps pleural and pericardial effusions, and a non-productive cough. The mortality among this special group of patients may reach 15%, but for the great mass of the population legionellas are virtually non-pathogenic. The public image of legionellosis and its reputation as a killer is undeserved: the media has focused attention on the tiny tip of a very small iceberg.

(b) Diagnosis and treatment

A diagnosis of legionellosis should be considered in patients with atypical pneumonia and in whom the predisposing factors apply. Legionellas may be isolated form sputum, bronchial washings, pleural or pericardial effusions, blood cultures, biopsies, or the lungs after death. The diagnosis may also be made serologically. Patients develop antibodies to an infecting legionella during the course of their illness, and these can be detected by appropriate tests.

In the laboratory legionellas appear to be sensitive to several antimicrobials that are ineffective in the treatment of patients. This is because they do not penetrate cells where the legionellas multiply. Erythromycin and rifampicin are the drugs of choice.

(c) Epidemiology and control

The recent emergence of legionellosis as a 'new' disease is probably related to the development and application of modern building techniques. Most big outbreaks of legionellosis have been associated with recently built or renovated hotels, offices and hospitals.

Legionellosis is an environmental infection that does not pass from person to person, so patients who suffer from it do not need to be isolated. It seems to be acquired by the inhalation of air that contains legionellas. For this to be possible tiny droplets must be generated from water in which legionellas have grown. These droplets must then be carried in currents of air to reach and be inhaled by people while the legionellas they contain remain alive and undamaged. It is possible to achieve this in the workings of some air-conditioning systems, or in the spray from shower heads or mixer taps. Additional sites that have been found to contain legionellas are chairside plumbing systems in dental departments and in humidifiers and respirators in intensive care units.

Because legionellosis is a disease that has appeared as a result of technical ingenuity, the control of it lies more in the province of engineers. Where water-cooled air-conditioning systems still exist they require very special attention. These, together with recirculating plumbing systems and humidification plants, need to be designed and maintained to reduce the possibility that legionellas can grow in them.

FURTHER READING

1.   Fraser, D.W., Tsai, T.R., Orenstein, W. *et al.* (1977) Legionnaires' disease. *New England Journal of Medicine*, **297**: 1183–97.
2.   See Appendix B.

13 The other Gram-negative rods

(Bordetella, Brucella, Haemophilus, Yersinia, Eikenella, Francisella, Gardnerella, Moraxella, Pasteurella)

13.1 INTRODUCTION

The nine bacterial genera listed in this residual group have little in common except that they are all Gram-negative bacilli that may cause human infections. A number are animal pathogens that may spread to humans, so the infections they cause are zoonoses. Four of the genera contain important pathogens for humans. Only a little need be said about the other five.

   Members of the more important group are, in alphabetic order, *Bordetella, Brucella, Haemophilus*, and the *Yersinia*. Many of the bacilli that fall into these genera are small in size when compared, for example, with the enterobacteria. Some are called **cocco-bacilli** because they are so short that they might be mistaken for cocci.

13.2 *BORDETELLA*

Bordetella pertussis is the cause of whooping-cough (pertussis). It is a small Gram-negative cocco-bacillus that tends to become larger and filamentous when it grows in the laboratory. *B. parapertussis* causes a disease indistinguishable from, though generally rather milder than, pertussis. The bordetella take two or three days to grow into visible colonies in the laboratory, even on specially enriched media.

13.2.1 *Bordetella pertussis*

(a) *Pathogenesis*

Whooping cough or pertussis is an acute disease of the respiratory tract. The bordetellas adhere to, and multiply on the cilia of the lining of the respiratory tract. The infection begins in a non-specific catarrhal fashion after an incubation period of seven to ten days. A cough, ordinary at the outset, develops a diagnostic paroxysmal quality over the next seven to 14 days. The paroxysms may continue with a variable frequency for one or two months. Each paroxysm consists of a series of very closely spaced, violent coughs, without intervening inhalations. The patient becomes distressed, and may be cyanosed. The paroxysm ends with a long inspiration through partially closed vocal cords that produces the characteristic 'whoop'. There may then be an expectoration of clear, tenacious mucus, perhaps with vomiting.

The disease is primarily of children, but atypical cases without the 'whoop' are seen in infants and adults. The mortality is low, but because those who die tend to be under one year old, it is a significant cause of death in this age group. Infants who survive may suffer permanent respiratory damage, for example, bronchiectasis. The disease is more common and severe in females.

Transmission is by direct contact with the respiratory discharges of a patient who is in the early catarrhal stage of the illness, before the diagnosis has become obvious. In households with several children the infection may be brought home by an older child, to infect infant brothers or sisters. By the time the characteristic 'whoop' has developed, the causative microbe has largely disappeared, and the patient is no longer infectious.

(b) Diagnosis and treatment

The laboratory diagnosis is made by culture of the organism from the naso-pharynx of the sufferer. A special per-nasal swab is used to collect the specimen (Figure 13.1).By the time the diagnosis is clinically obvious, cultures are often negative. Although the microbes are sensitive to several anti-microbials, including erythromycin, these are only of value in the early, non-specific stage of the disease, that is, before the diagnosis is likely to be made. Once the characteristic cough has developed the causative organisms have gone, so antimicrobials are without effect. The infection is accompanied by an increase in the number of lymphocytes in the blood, though this change may not be seen in infants.

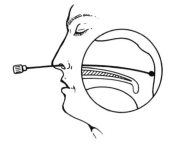

Figure 13.1 The special per-nasal swab used for the collection of specimens for culture in whooping cough. The stalk of the swab is made of wire, and the tip of something other than cotton wool, which if used seems to reduce the chances of a successful isolation.

(c) Control

Vaccines that contain whole, killed *B. pertussis* give good protection. They are usually mixed with diphtheria and tetanus antigens to form the triple vaccine DTP. Pertussis vaccination has been accompanied by a very low incidence of adverse neurological reactions (Chapter 31). Modified vaccines have been produced that seem to reduce the risk. In some countries it is advised that infants with a history of convulsions should be given DT and not DTP. As children under one year of age suffer most severely from pertussis, the vaccine is used as early as possible. Unfortunately infants under six to eight weeks old do not respond to it. Adults whose immunity has waned after vaccination in childhood may be reinfected if they come into contact with a child or another adult with whooping cough. The disease suffered by these adults may appear to be no more than a 'cold', but it is infectious to others. This is important in hospitals, particularly for staff who come into contact with infants.

13.3 BRUCELLA

(a) Pathogenesis

The brucella are small Gram-negative cocco-bacilli that multiply inside the cells of animals and humans to cause brucellosis. In animals the infection may be mild and pass unnoticed, but pregnant animals (cattle, pigs, sheep and goats) tend to abort. Brucellas multiply freely in the placentas of these animals because they contain the substance erythritol, which encourages their growth. There is no erythritol in human placentas so abortion is not a feature of infection among women. The disease is essentially one of animals. Humans are involved by accident, and they are 'dead-end hosts' because they play no part in the continued existence of the bacterium in nature.

After an incubation period of 1–6 weeks, human brucellosis begins either acutely or insidiously with a fever that is characteristically irregular (an alternative name for brucellosis is undulant fever) with headaches, weakness, profuse sweating, chills, pains in the joints and mental depression. This can go on for days, weeks or months. The spleen is commonly enlarged. Recovery is usual, but disability may be pronounced. Some sufferers enter a chronic phase, with repeated febrile attacks accompanied by a feeling of weakness and severe mental depression.

The various brucella species are associated with particular animal hosts, as shown in Table 13.1. For humans *Brucella melitensis* is the most virulent. The four species can be distinguished by laboratory tests.

Table 13.1 The association between the species of brucella and the animal hosts in which they are more commonly found

Brucella spp.	Animal hosts
B. abortus	cattle
B. melitensis	sheep, goats
B. suis	pigs
B. canis	dogs

(b) Diagnosis and treatment

Brucellas are isolated from blood cultures taken from patients in an acute febrile episode (and only then), or from pus in rare cases of local brucella sepsis. *B. melitensis* is more often isolated from cultures because, as the most pathogenic variety, they are present in the largest numbers. The diagnosis is often made serologically. For treatment a prolonged course of tetracycline is used, sometimes with added streptomycin.

(c) Epidemiology and control

Brucellosis is found worldwide. The different brucella species predominate in a given locality according to the distribution of their animal hosts. The disease is common among people who tend infected herds of animals, so nomadic herdsmen, farmers, butchers and veterinary surgeons are often involved. When an animal with brucellosis aborts, the placenta and amniotic fluid contain brucellas in large numbers, and people present at the time are heavily exposed to infection. The disease can also be transmitted by drinking unpasteurized milk from infected animals, or eating soft cheese made from such milk. It does not spread from person to person, and is controlled by eradicating the disease in domestic animals. This has been achieved in a few countries, but it is a slow and expensive process. Animals can be vaccinated against the disease.

13.4 HAEMOPHILUS

Haemophilus influenzae is the most important human pathogen in this genus. *H. ducreyi* causes a sexually transmitted disease. Some other members of the genus are found as commensals in the upper respiratory tract.

13.4.1 H. influenzae

This bacterium is characteristically variable in shape, appearing as a coccobacillus, or as long filaments with occasional larger spherical forms. When fully developed it bears a polysaccharide capsule of one of six antigenic types, labelled a–f. *H. influenzae* type b (Hib) is the main human pathogen. The capsule is lost on culture in the laboratory, and non-capsulated variants are commonly found as commensals in the upper respiratory tract.

   H. influenzae grows poorly or not at all on ordinary media, as it has an absolute requirement for two vitamin-like growth factors. Ordinary blood agar is deficient in one of these so *H. influenzae* grows more luxuriantly round colonies of other microbes present at the same time that produce an abundance

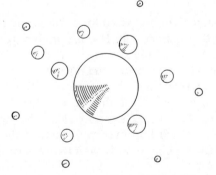

Figure 13.2 Satellitism, with colonies of *H. influenzae* that grow larger as they approach the colony of *Staph. aureus* at the centre. *Staph. aureus* produces an excess of a vitamin called 'V factor'. The haemophilus needs this but cannot make it for itself, and it is absent from the culture medium. The other vitamin it needs ('X' factor) is provided in the blood added to the medium.

of it. This is called 'satellitism' (Figure 13.2). Various members of the genus *Haemophilus* are distinguished from each other by their different requirements for these two factors.

(a) Pathogenesis

In its pathogenic capsulated form *H. influenzae* causes suppurative respiratory disease in children or, less commonly, in adults. It may cause laryngotracheobronchitis, sinusitis, otitis media or conjunctivitis, or, more rarely, epiglottitis or meningitis. Bacteraemia or septicaemia may complicate (or in the case of meningitis precede) any of these. When an individual meets the microbe for the first time, usually in childhood, it causes an upper respiratory infection that is usually mild, but which produces immunity. In a few individuals this initial infection develops into one of the others listed, of which epiglottitis and meningitis are life-threatening. In most parts of the world *H. influenzae* is a common cause of childhood meningitis. In epiglottitis the swelling of the larynx and epiglottis may be so intense that the trachea is completely obstructed and an emergency tracheostomy is required to save the patient's life.

(b) Diagnosis and treatment

Diagnosis is by culture of appropriate clinical specimens. In meningitis the examination of a Gram-stained smear of CSF may allow a quick diagnosis to be made, unless the patient has been treated with an antimicrobial drug prior to lumbar puncture.

Treatment is complicated by the appearance of antimicrobial resistance among *H. influenzae*. Some of them now produce the enzyme beta-lactamase, and because of this up to 30% may be resistant to ampicillin. Chloramphenicol may be used instead, but strains may be resistant to this as well. So far cefotaxime has proved an effective alternative.

(c) Control

Prophylaxis with rifampicin is advised for children and others who have been in close contact with a patient with Hib meningitis, particularly when they belong to the same family. This is because there is a 5% attack rate among children under two who have been in contact with another case, compared with an attack rate among children in general of 0.4%. A vaccine is made from the polysaccharide present in the capsule. On its own this material is poorly antigenic in the under two-year-olds who are most at risk. Its antigenicity has been much improved by attaching the antigen to another, larger molecule. Hib vaccine is now widely used for the immunization of infants (Chapter 31).

13.4.2 *H. ducreyi*

This is the cause of the sexually transmitted disease chancroid or soft chancre. The latter name distinguishes it from the hard chancre of syphilis. Three to five days after exposure to an infected partner single or multiple soft painful necrotic ulcers appear on the genitals with enlargement of the local lymph nodes. Later these may break down and suppurate. The disease is more common in seaports in tropical or subtropical areas. Microscopy reveals the causative organism in stained smears made from exudate, pus or a biopsy. They appear as small Gram-negative rods that are often arranged in chains. They can be grown in the laboratory. Treatment is with tetracycline, cotrimoxazole or erythromycin.

13.5 YERSINIA

The yersinias are animal pathogens that sometimes spread to humans to cause disease. *Yersinia pestis* (formerly called *Pasteurella pestis*) is the cause of plague while *Y. enterocolitica* and *Y. pseudotuberculosis* attack the gastrointestinal tract.

13.5.1 *Y. pestis*

(a) Pathogenesis

Plague exists in the wild in many parts of the world. It circulates among rodents in particular, and passes between them as a result of the bites of fleas, ticks or lice, or when an infected animal is eaten by another. This 'sylvatic' plague is a relatively mild disease among the animals most commonly involved.

If conditions are right *Y. pestis* can escape from this sylvatic maintenance cycle to reach 'domestic' rats that live in and around human habitations. Among these the bacillus is spread by fleas, and both rats and fleas suffer an infection with a high mortality. When the rats die their fleas (which survive longer) are driven to seek alternative sources of blood for food, and humans may then be involved. If the infestation with rats was heavy, a great many infected fleas are available to attack a human population. The consequence is an epidemic of what was once called the 'black death'. Between 1347 and 1351 this is estimated to have killed between a quarter and a half of the population of Europe. Plague visited Europe again in waves between 1613 and 1666, once more with a heavy mortality (Box 13.1).

Box 13.1 A plaguey piper

Robert Browning's poem 'the pied piper of Hamelin' is an allegory. The 'pied piper' was the plague that first rid the town of the rats by which it was overrun, then killed the children who vanished into a cavern (their graves) on the Koppelberg hill outside the town.

When plague kills children rather than adults the latin description is 'pestis puerorum'. This pattern of mortality is characteristic of the second or third waves of plague in a given population. In the first wave people of all ages are killed, but many of the survivors are immune. In the course of the·next 10 or 15 years the number of non-immune children grow to form a susceptible population that is selectively attacked when the next epidemic wave appears (Wolfers, 1965).

Sporadic cases of plague are found among people occupationally or recreationally exposed to animals or their insect parasites in areas where sylvatic plague exists. At present these places include parts of the USA, South America, Africa, and SE Asia. Other parts of the world have been involved in the past, and may be again in the future.

Y. pestis is a Gram-negative cocco-bacillus. It grows best at 27°C, which suits it for the part of its life it spends in cold-blooded insects. In an infection a powerful necrotizing toxin is produced that usually overcomes attempts by the body to localize it. Areas of haemorrhagic necrosis are produced that contain large numbers of the bacilli. In fulminating infections there is little inflammatory response, though when they are less overwhelming, pus may be produced. The four main types of plague are outlined in Box 13.2.

Box 13.2 The four types of plague

When *Y. pestis* is inoculated into the skin by the bite of an infected insect the result can be **bubonic plague**. The microbe multiplies locally and spreads to the nearest lymph glands. Bites are often on the leg so the glands are in the groin. These swell to form the tender **bubo**, from which the bacteria extend into the blood to cause septicaemia. If untreated the mortality is 60–90%.

Septicaemic plague develops when *Y. pestis* reaches the circulation without the involvement of lymph glands. This form of the disease is rapidly and uniformly fatal.

Pneumonic plague spreads from person to person in crowded situations. The lungs are widely involved and pneumonia soon turns into septicaemia. This form of the disease is very infectious and highly lethal.

Pestis minor is a mild disease. Septicaemia does not develop and the sufferer becomes immune. In an epidemic this is probably the most common form of plague.

(b) Diagnosis, treatment and control

Y. pestis can be identified microscopically in smears made from material aspirated from buboes with a syringe and needle. It can be grown in the laboratory from this fluid, or from blood or sputum, according to the clinical picture. Streptomycin and tetracycline have been used in treatment.

A plague vaccine is available. The avoidance of wild rodents in areas of sylvatic plague, and the control of 'domestic' rats are important general public health measures. International control is by the enforcement of rules designed to prevent rats from travelling from country to country on board ships, trains or aircraft.

13.5.2 *Y. enterocolitica* and *Y. pseudotuberculosis*

These related bacteria cause infections of the gastrointestinal canal. In most cases the disease is mild and self-limiting.

In some parts of the world *Y. enterocolitica* is an important cause of enterocolitis, often associated with enlargement of the abdominal lymph nodes. Yersiniosis is a febrile diarrhoea acquired when food is eaten that has come from, or has been contaminated by, an infected or carrier animal. Abdominal pain due to the lymphadenitis may be severe and mimic appendicitis, leading to operation. The condition is commonly associated with a bacteraemia. Infection with *Y. pseudotuberculosis* follows the same pattern, particularly in young children.

Yersinias may be cultured from faeces, blood, or from specimens taken at operation, and from the food that was their source. Antibodies to yersinias may also be detected in the serum of patients who have been infected. The yersinias share with the listeria an ability to continue to multiply in a refrigerator. A person may donate blood while suffering from a mild attack of yersiniosis, but the blood collected contains so few yersinias that they are undetectable by normal culture. After storage for up to three weeks in a refrigerator the few yersinias have been converted into millions. A person who receives a transfusion of blood in this condition will not welcome the added insult.

13.6 THE REMAINING GRAM-NEGATIVE RODS

13.6.1 *Eikenella corrodens*

In the laboratory this bacillus produces colonies that dissolve (corrode) the agar on which they grow. They are members of the normal upper respiratory flora, and may be found with other microbes in wounds and abscesses, and occasionally in meningitis and endocarditis.

13.6.2 *Francisella tularensis*

This is a parasite of rodents (rabbits, hares, lemmings, etc.) among which it is spread by fleas, ticks and other arthropods. The infection is called tularaemia. Human cases arise by contact with infected animals when they are trapped, skinned or eaten, or result from the bite of an insect vector. The disease is endemic in North America, northern Europe, the USSR and Japan.

The disease presents in a variety of ways, depending on the route of infection and the virulence of the strain concerned. If it is introduced through the skin an ulcer develops at the site, with swelling of the local lymph glands. The inhalation of infected material leads to a pneumonia, and its ingestion to pharyngitis, abdominal pain, diarrhoea and vomiting. Any of these local infections may progress to a fatal septicaemia. Treatment is with streptomycin or chloramphenicol.

F. tularensis is a small Gram-negative rod that requires enriched media for culture in the laboratory.

13.6.3 *Gardnerella vaginalis*

This is a normal inhabitant of the vagina, but it multiplies abnormally when there is a vaginitis. There is argument about its pathogenicity but on balance it seems that the organism contributes to rather than causes the infection. 'Clue cells' are vaginal epithelial cells seen in stained smears to be covered with diffrent kinds of bacteria many of which are small Gram-negative rods that are thought to be *G. vaginalis*. Clue cells are found in cases of 'bacterial vaginosis' or non-specific vaginitis. Treatment with metronidazole usually relieves the condition .

13.6.4 *Moraxella* spp.

These short thick Gram-negative diplobacilli may be difficult to grow in the laboratory. *M. lacunata* can cause a purulent conjunctivitis or, rarely, a generalized infection. They are normal inhabitants of the upper respiratory and genital tracts. In the latter site they may be mistaken for *Neisseria gonorrhoeae*.

13.6.5 *Pasteurella multocida*

P. multocida is carried as a part of the normal flora of the mouth by many animals among which it may also cause a haemorrhagic septicaemia. It is transmitted to humans by animal bites or scratches, to cause acutely inflamed wounds. Septicaemia and meningitis are rare complications. *P. multocida* can also cause chest infections. Treatment is with penicillin.

   P. multocida is a small Gram-negative rod. Its presence can be confirmed by the culture of material taken from wounds, or of sputum.

FURTHER READING

1.   Wolfers, D. (1965) A plaguey piper. *Lancet*, **1**, 756–7.
2.   See Appendix B.

The spiral bacteria $\boxed{\textbf{14}}$
(Treponema, Leptospira, Borrelia, Spirillum)

14.1 INTRODUCTION

The spiral bacteria or **spirochaetes** are a collection of free-living, parasitic or pathogenic bacteria of an unusual shape. Those of importance in medical microbiology are listed in Table 14.1, with the diseases they cause.

Table 14.1 Spiral bacteria of importance in human infections

Species	Diseases
Treponema	
pallidum	syphilis
pertenue	yaws
carateum	pinta
Leptospira interrogans (many varieties)	leptospirosis
Borrelia	
recurrentis	louse and tick borne relapsing fevers
burgdorferi	Lyme disease
vincenti	Vincent's angina, etc.
Spirillum minor	rat-bite fever

Spirochaetes are motile, thin-walled, flexible bacteria. Fibrils run longitudinally through them from end to end to form an axial filament. The bacterium and filament twist round each other to give spirochaetes their characteristic

---

Box 14.1 Differences between the spirochaetes

Treponemes have fine, regular spirals. They do not stain with Gram's or other simple stains. For most purposes dark ground microscopy is used to make them visible. They have not been cultured in the laboratory, but they can cause infections in some animals.

    Leptospires have regular spirals finer than treponemes, otherwise they the same except that they can grow in special media in the laboratory.

    Borrelia have coarse irregular spirals. They stain readily with ordinary dyes, and can be grown in the laboratory.

    (See also Figure 5.1)

---

shapes. Although *Spirillum minor* is also spiral in shape it has a different structure and it is not a true spirochaete. Box 14.1 and Figure 5.1 show how the different spirochaetes may be made visible, and how they are distinguished from each other.

14.2   THE TREPONEMES

(a)   *Pathogenesis*

The diseases caused by the named species of the genus *Treponema* (Table 14.1) are collectively known as the treponematoses. *T. pallidum* causes syphilis, which may be venereal, with sexual transmission (Box 14.2), or non-venereal ('endemic'). In the latter form the infection is usually acquired by children brought up in crowded nomadic situations, or under very unhygienic conditions. Transmission is often oral, and shared drinking vessels may play a part in this. *T. pertenue* causes the tropical non-venereal treponematosis called yaws. This is found in hot damp climates in parts of Africa, Asia, Latin America and the South Pacific among people living in unhygienic conditions. Transmission is by close, predominantly non-sexual contact. *T. carateum* is the cause of pinta, found in the tropical parts of America, where it is the New World counterpart of yaws.

Box 14.2 The great pox

Syphilis used to be called this to distinguish it from smallpox. Not all historians agree that Christopher Columbus brought it back when he returned from the New World, but there is no doubt that the disease cast a new and sinister shadow over Europe from the end of the fifteenth century. Henry VIII, Edward VI and 'Bloody' Mary I of England, Louis XIV and Henry III of France and Ivan the Terrible of Russia are all thought to have suffered from it. The French called it the 'Neapolitan evil', and the Italians called it the 'French evil'. Vasco da Gama carried it to India, in 1498, and the 'French pox' arrived in London in 1503. In 1767 the surgeon John Hunter inoculated himself with what he thought was gonococcal pus, and acquired syphilis as well. The diseases were separated by Benjamin Bell in 1793. In 1905 the spirochaete was identified by Fritz Schaudinn and P. Hoffman. Treatment with poisonous mercury was abandoned in 1910 when Paul Erlich and Sahachiro Hata introduced the arsenical antimicrobial, Salvarsan.

Although these diseases vary widely in their outward manifestations similar infectious and pathological processes underly all of them. The differences are due to characteristic variations in the intensity with which the different treponemes attack various organs and tissues.

   After exposure to infection there is an incubation period of three or four weeks. The treponemes multiply at the site of innoculation, usually on the genitalia in venereal syphilis and elsewhere on the body for the other treponematoses. A primary lesion then develops at this site and at the same time the treponemes begin to spread through the rest of the body. The primary stage begins with the appearance of a papule, with enlargement of the local lymph glands. In syphilis the papule develops into a painless ulcer, with a clean hard base, to form the typical, usually single, 'primary chancre'. In the other treponematoses there may be several primary sores that enlarge but do not always ulcerate. If not treated these primary lesions may take several months to heal and during this time the individual can transmit the infection to others.

After this the individual enters the secondary stage of the disease. In syphilis lesions of two general types appear: blotchy rashes anywhere on the body, and soft, pale wart-like lesions (codylomas) in the urogenital region and the mouth. Treponemes are easily found in either of these, and the individual is highly infectious. In the other treponematoses the secondary lesions have appearances that are typical of the diseases concerned. These heal after a period of months or years, and the secondary stage terminates in a latent period, during which the activity of the treponemes is suppressed, or they are eradicated. In 60–70% of cases of syphilis this remission is permanent.

The other 30–40% of patients enter the tertiary stage. Once more lesions of two basic types appear, though these are distributed in different tissues according to which treponeme is responsible (Table 14.2). One type of lesion is the gumma, a chronic inflammatory reaction with central necrosis that affects skin, bones, the liver and other viscera. The second is an inflammatory proliferation of the linings of small arteries that leads to blockage and damage due to lack of blood downstream ('obliterative endarteritis'). This is the underlying cause of the pathology typical of tertiary syphilis.

Table 14.2 The distribution and intensity of lesions in the tertiary stages of the different treponematoses

Lesions of the	Venereal syphilis	Endemic syphilis	Yaws	Pinta
Placenta	+++*	−	−	−
Viscera	+++	−	−	−
CVS	+++	+	−	−
CNS	+++	+	−	−
Bones	++	++	+++	−
Skin	++	++	+++	+++

CVS: cardiovascular system
CNS: central nervous system
−: no lesions
+ to +++: increasing number and severity of lesions
*An infection of the placenta spreads to the fetus. This may cause fetal death or the birth of a congenital syphilitic who will develop a variety of typical signs and symptoms in childhood. Treatment of the mother during pregnancy can prevent this, so all expectant mothers should be tested as a routine for serological evidence of syphilis

The arteries that nourish the walls of the aorta and other great vessels are often involved and their vital elastic tissues are destroyed. Without their elasticity the walls bulge under the pressure of the arterial blood inside them. Permanent swelling or aneurysm develops that tends to increase in size, and may eventually leak or burst. Obliterative endarteritis in the central nervous system (CNS) leads to a variety of mental and neurological changes such as tabes dorsalis and general paralysis of the insane. Treponemes are very scanty in these lesions, which seem to be caused by an immunological reaction to their presence. Patients in the tertiary stage of syphilis are not infectious and do not need to be isolated.

Patients with treponemal infections develop two kinds of antibody. One of them is specific for the infecting treponeme, the other is a curiosity in that it can react with fatty extracts of perfectly normal mammalian tissues. This 'reagin' antibody is probably an auto-antibody produced against host tissues damaged by treponemes. As all the treponemes are identical antigenically, serological tests cannot distinguish between any of the diseases they cause. The distinction is made from the history and a clinical examination.

(b) Diagnosis and treatment

Treponemes can be seen in fluid expressed from primary or secondary lesions when it is examined by dark-ground microscopy. When the result of such microscopy coincides with the clinical evidence, a presumptive diagnosis is made and treatment is started. It is important to note that non-pathogenic treponemes microscopically indistinguishable from *T. pallidum* may be found among the normal flora of, for example, the mouth.

The mainstay of diagnosis is a serological examination for antibodies. Very many different tests are available because practically every possible serological technique for the demonstration of antigen-antibody reactions has at one time or another been applied to the diagnosis of syphilis. The matter is further complicated by the existence of the two different antibodies, for which separate tests must be done.

It is customary to begin with a test that detects reagin. The ones commonly employed are the Venereal Diseases Research Laboratory (VDRL) or the rapid plasma reagin (RPR) tests. Positive results are confirmed using a treponeme-specific test like the fluorescent treponemal antibody (FTA) or the *T. pallidum* haemagglutination (TPHA) tests. Both kinds of test become positive very early in the primary stage of infection, and remain positive until the patient is treated or is cured spontaneously. Reagin-type antibodies then begin to disappear, though tests for them may take a year or more to become negative. The specfic tests remain positive for longer, and in patients who had long-standing infections prior to treatment they never become negative. A patient successfully treated for tertiary syphilis who still has signs and symptoms due to permanent tissue destruction may test negative for reagin and positive for specific antibody. A positive serological test on a specimen of CSF is taken as strong presumptive evidence of a treponemal infection of the CNS.

(c) Epidemiology and control

When socio-economic conditions improve in the society concerned, non-venereal treponemal infections disappear. In some parts of the world penicillin has been used in mass campaigns to reduce the prevalence of yaws and pinta. The control of venereal syphilis depends on the provision of services that combine the confidential treatment of cases of the infection with a search for their contacts. In theory this allows for the treatment of an expanding circle of individuals with syphilis and a reduction in the pool of infection in the community. This approach has not been an outstanding success.

14.3 LEPTOSPIROSIS

(a) Pathogenesis

Leptospirosis is an infection primarily of animals caused by the spiral parasitic bacterium, *Leptospira interrogans*. This microbe, named for an occasional resemblance to a question-mark, spreads readily in many different kinds of animal, among which it usually causes little or no disease. Humans, in whom it can cause severe illness, are infected by accident. Leptospires survive for long periods in natural water, provided it is neither too salt nor too acid. *L. interrogans*, the pathogenic species, has a counterpart in *L. biflexa*, a free-living non-pathogenic form.

L. *interrogans* is subdivided into a large number of serologically distinguishable varieties or **serovars**. Each of these tends to be associated with a particular animal species, and perhaps a special geographical location. Most animal-leptospire relationships are stable and long-term. Initial, perhaps subclinical, illnesses are followed by colonization of the hosts' kidneys. These suffer no damage as leptospires spill out into their urine. The urine contaminates the environment to provide routes by which leptospires reach new hosts. The most important of these environmental routes is water. When humans enter an environment occupied by infected animals they risk infection because leptospires can penetrate the intact skin, or they may consume food or water that contains them. The animal hosts most commonly infected are wild rodents, but domestic animals (cattle, dogs, cats, etc.) are also involved.

In humans the infection is often mild or subclinical. More serious forms of the disease usually present in two stages. The organs involved and the severity of the illness depend on the serovar of the leptospire involved. The general pattern is that, after an incubation period of about 7–12 days, there is an initial phase when leptospires circulate in the blood. This is accompanied by an acute feverish illness, influenzal in style, perhaps with a rash. In many cases the disease resolves at this stage but in a few individuals more ominous signs appear after about a week. These may suggest severe involvement of the liver or kidneys, or both. Jaundice and a diminished secretion of urine accompanied by haemorrhages indicate the development of the life-threatening form of the infection called Weil's disease. This classical but by no means the most common form of leptospirosis is caused by *L. interrogans* var *icterohaemorrhagiae*, a parasite of rats. Other cases are marked by an acute meningitis that appears in the second week. Because human urine is commonly acid, particularly when there is a fever, leptospires do not easily survive in the human kidneys. Chronic renal infection and person-to-person spread of the disease is rarely a problem.

(b) Diagnosis

Blood cultures may be positive during the early acute phase of the infection. The special medium required is not often used because at the stage when leptospires are present in the blood there is little to suggest the diagnosis. This is usually made later, serologically. The tests are of two kinds. One gives a clinically useful diagnosis of 'leptospirosis'. The other is more specific and can distinguish between infections due to the different leptospire serovars. This is of importance epidemiologically, as it gives a clue to the source of the infection.

(c) Treatment and control

By the time the diagnosis of leptospirosis is made, the acute septicaemic stage of the illness, when appropriate antimicrobials might be useful, is usually over. Late treatment with penicillin or tetracycline is of limited benefit. Patients in the early stages of infection when abundant leptospires are present in their tissues very often suffer a sharp though brief deterioration of their condition about two hours after the initiation of treatment. This is called the 'Jarisch-Herxheimer reaction' (Box 14.3).

Box 14.3 The Jarisch-Herxheimer reaction

This reaction is seen when one of the treponematoses is treated with an antimicrobial like penicillin in a phase of the infection when treponemes are present in large numbers, as in secondary syphilis. The same thing happens in relapsing fever, early in leptospirosis or in Lyme disease.

One or two hours after the first dose of the drug has been given, the patient suffers an acute febrile reaction with headache, muscle pains, flushing and a fall in blood pressure that may last 24 hours. The reaction is due to the rapid destruction of large numbers of the spiral microbes with the release of quantities of endotoxin (Chapter 3) or of antigens that are involved in extensive antigen-antibody reactions.

Because leptospirosis is acquired from infected animals or their urine the disease may be difficult to avoid. Farmers and others regularly exposed to animals can try to keep their charges (and so themselves) free of infection by the control of wild rodents, and if necessary their animals can be vaccinated against the disease. For others avoidance of potentially contaminated fresh water, or if this is impossible the wearing of waterproof protection, is the only answer. Sea water is quite safe.

14.4 THE BORRELIOSES

The borrelia and the diseases they cause are listed in Box 14.4.

Box 14.4 Borrelia and the borrelioses

General infections
- with *B. recurrentis*:

  1. Louse-borne epidemic relapsing fever
  2. Soft-tick-borne endemic relapsing fever

- with *B. burgdorferi*:
  Hard-tick-borne Lyme disease

Superficial infections
- with *B. vincenti*:
  Vincent's angina, tropical ulcer, etc.

14.4.1 Relapsing fever

(a) Pathogenicity

In relapsing fever, febrile episodes that last several days alternate with short afebrile periods. The epidemic louse-borne disease is more severe with fewer relapses, and borrelia are easy to find in the blood. Endemic tick-borne relapsing fever is less severe and borrelias are present in the blood in smaller numbers, but there are more (up to ten) relapses. In either case a rash is common in the first febrile episode. The relapsing pattern of the disease seems to be due to the ability of borrelia to undergo minor antigenic changes between successive bouts of fever, so for a time they stay one step ahead of eradication by the body's defences.

The endemic variety of the infection is widely distributed through tropical Africa, the Middle East, India, and parts of Central Asia and North and South America. The disease is one of rodents, among which it is spread by soft ticks, which can also pass the borrelia to their own offspring, from one generation to another. Humans are involved when they move through an area inhabited by the ticks. The organism is transferred by bite, or by the contamination of a bite or scratch by the body fluids of the tick.

Epidemic louse-borne relapsing fever appears in conditions of overcrowding and malnutrition such as result from a disaster of natural or human origin. When lice are introduced into such a situation they proliferate. If some of them take blood meals from individuals with borrelia in their blood, the lice are infected. The lice concerned are strictly human parasites so when they seek new hosts on whom to feed they transmit their borrelia to a widening circle of people. Borrelia are not found in the mouth parts of infected lice. When they are crushed by scratching, their body fluids and the spirochaetes these contain are rubbed into the bites, or into the scratches that result from the irritation. In this way a man-louse-man cycle of transmission is established and in the right circumstances an epidemic of relapsing fever is the result (World, 1993).

(b)   Diagnosis, treatment and control

Borrelia are present in the blood of sufferers during febrile episodes. They can be found in blood films stained by Giemsa's method, or in wet films examined by dark-ground microscopy. Treatment is with tetracycline, erythromycin or penicillin. The systemic borrelioses can be prevented by the avoidance of exposure to ticks, or the control of human lice (Chapter 27).

14.4.2   Lyme disease

This recently-described zoonosis is caused by *B . burgdorferi*. The spirochaete is maintained in nature by a cycle that involves various mammals and some of their bloodsucking hard-tick parasites. Humans who invade a territory in which this cycle operates may be bitten by an infected tick. Between three and 21 days later the individual develops a red spot at the site of the bite and this extends outwards to form the large and expanding annular rash called erythema chronicum migrans. This is accompanied by fever, headache and a stiff neck that lasts for several weeks. At any time in the next two years arthritis and a chronic meningitis may develop. The diagnosis is made by detecting antibody to the borrelia in the serum of the patient. Treatment is with tetracycline or penicillin. Control is by avoiding areas inhabited by infected wild animals and their ticks.

14.4.3   The superficial borrelioses

B. vincenti is a usually harmless inhabitant of the mouth. In conditions of malnutrition or immunodeficiency compounded by injury or other infections, borrelias invade areas of damaged skin or mucous membrane. In these they set up a spreading, foul-smelling gangrenous necrosis. The borrelia are always accompanied by other bacteria, among which the Gram-negative anaerobe *Fusobacterium fusiformis* is the most common (Chapter 8). The lesions that result have many names. When they involve the mouth it may be called Vincent's angina (Vincent's stomatitis) or trench mouth, or elsewhere noma or tropical ulcer. The diagnosis is made by microscopy. Stained smears from lesions show the presence of the participating microbes. Treatment is with penicillin, plus attention to the underlying cause.

14.5   *SPIRILLUM MINOR*

S. minor does not have the anatomy of a true spirochaete. It is a small rigid motile spiral microbe that is found worldwide as a parasite of rats. It is one of the two causes of rat-bite fever. (The other is *Streptobacillus moniliformis*.) A lesion develops at the site of a rat-bite, accompanied by enlargement of the local lymph nodes together with a relapsing type of fever and a rash that fluctuates in intensity as the temperature goes up and down. The diagnosis is made when the spirillum is found in local lesions by dark-ground microscopy, or in stained smears of blood. It may be isolated by the inoculation of specimens into guinea-pigs, in which a fatal disease is produced. Treatment is with penicillin.

FURTHER READING

1. World, M.J. (1993) Pestilence, war, and lice. *Lancet*, **342**, 1192.
2. See Appendix B.

15

The atypical bacteria
(Mycoplasma, Ureaplasma, Chlamydia, Rickettsia)

These microbes are atypical because though they share major properties with other bacteria they differ from them in certain important respects.

15.1 THE MYCOPLASMAS AND UREAPLASMAS

These are the smallest and simplest free-living bacteria that can multiply without the aid of other living cells. They lack the peptidoglycan that confers rigidity on conventional bacteria, so they vary in shape and size. As Gram's stain acts on cell walls they are neither Gram-positive nor Gram-negative. To make them visible for microscopy a stain such as Giemsa's can be used (Box 15.1).

Box 15.1 Giemsa's stain

This stain, invented by a German of the same name, is one of the Romanowsky group that combine various red and blue dyes in different mixtures. They are widely used for blood films, in which they also stain malaria and other parasites. They may be used to stain microbes that do not react to Gram's stain in smears of tissue fluid or pus.

Each species of mycoplasma is a parasite of one kind of host, human, animal, insect or plant. A growing number of species of these genera have been incriminated as human pathogens with the genera *Mycoplasma* and *Ureaplasma* of importance. The best known is *M. pneumoniae*, the cause of mycoplasmal, or 'primary' atypical pneumonia. Special media and techniques are necessary to cultivate mycoplasmas in the laboratory, and they grow rather slowly.

15.1.1 *M. pneumoniae*

The primary site of multiplication of *M. pneumoniae* is usually the pharynx, where it settles following inhalation or ingestion. After an incubation period of 14–21 days a mild, usually self-limiting pharyngitis develops. In some people the infection extends to cause bronchitis and in about a third of infections this becomes an atypical pneumonia. When this happens an X-ray examination usually shows that the disease is more extensive than the physical signs suggest. The patient may be quite ill, though the infection is rarely fatal. Mycoplasmal pneumonia is more common in children over five, teenagers and young adults. Mycoplasmas spread from person to person, mainly by close contact. Because many of the cases are very mild a chain of infection is rarely identified.

(a) Diagnosis and treatment

M. pneumoniae can be cultured from the sputum though because of technical difficulty this is rarely attempted. The diagnosis is usually made serologically. As the specific antibodies develop in an infection others called 'cold agglutinins' often appear. These agglutinate human group 'O' erythrocytes in the refrigerator, but not at 37°C. Penicillins and cephalosporins are of no use in the treatment of infections due to *M. pneumoniae* because there is no cell wall to be attacked. Tetracycline and erythromycin are effective.

15.1.2 The other mycoplasmas and ureaplasmas

M. hominis and *U. urealyticum* are thought to be the causes of some cases of non-gonococcal urethritis in males, and urethritis, cervicitis, salpingitis and chronic pelvic inflammatory disease in females. Equally it has to be said that a significant number of apparently healthy, sexually active people carry these organisms in their genital tracts, with no evidence of pathology. Much remains to be learnt about the activity of these tiny bacteria.

15.2 THE CHLAMYDIAS

Chlamydias can only multiply in living cells. Because of this they were originally classified as viruses. Although they are now known to be bacteria, virological methods have to be used to grow them in the laboratory.

Chlamydias cause infections in a wide variety of animals and birds, and in humans. The genus *Chlamydia* contains two species, *C. psittaci* and *C. trachomatis*. The normal hosts of *C. psittaci* are animals and birds. When humans are involved it is by accident. By contrast *C. trachomatis* is an entirely human pathogen. A newly discovered chlamydia, *C. pneumoniae*, is related to *C. psittaci*, but is also primarily a human pathogen.

Mature chlamydias resemble very small Gram-negative bacteria. They contain DNA and RNA, have rigid cell walls and multiply by binary fission. They possess an incomplete range of metabolic enzymes and to make up the deficiency they multiply inside other living cells. When a chlamydia has been taken into a cell by phagocytosis it changes into a reproductive form that grows and then divides into multiple copies of itself. The process may be followed under a microscope, and the mass of developing chlamydias, called an inclusion body, can be stained with Giemsa's stain.

All chlamydias possess a common group antigen, plus others that divide the genus into species and varieties. These divisions and subdivisions and the human infections they cause are shown in Box 15.2.

Box 15.2 Human chlamydial diseases

C. psittaci	Psittacosis
C. pneumoniae	Pneumonia
C. trachomatis	
First group	Trachoma
Second group	Non-gonococcal urethritis and commplications (Table 15.1)
Third group	Lymphogranuloma venereum

15.2.1 *C. psittaci*

(a) Pathogenesis

This chlamydia is the cause of psittacosis (or ornithosis), a disease principally of birds, although other animals may be involved. Among birds the infection is widespread, and usually latent. Stress due to capture and transportation, migratory flights or ovulation and the feeding of young may cause birds to increase the rate at which they shed chlamydias in their droppings and nasal secretions. In this way nestlings are infected while still immunologically immature and under the protection of maternal antibody. This, passed to them through the yolk of the egg, explains why the infection in birds is usually asymptomatic, and why chlamydias persist from generation to generation. Capture and transportation causes illness due to stress. Newly caged birds have in the past caused outbreaks of the infection among humans exposed to them. The disease was at first called psittacosis because of its association with the importation of exotic psittacine birds, particularly parrots. It is now clear that many other types of bird and animal may be involved.

Infection is transmitted by the inhalation of dust that contains the dried secretions or excretions of birds that shed the virus. Outbreaks of the disease have been described in people occupationally exposed to flocks of chickens, turkeys, ducks and geese, or to feathers from them. Household pet birds including budgerigars can be the source of infection. Person-to-person transmission is very rare.

The severity of the disease varies widely. This may reflect differences in the vigour of the host response and the size of the infecting dose. An apparently fit bird probably sheds few if any chlamydias, while a sick one may scatter them in large numbers. The disease has an acute generalized onset with fever, headache and muscle pains. This may develop into a lower respiratory illness with patchy and often extensive pneumonia. Cough may be mild or absent, and enlargement of the spleen is common. In some outbreaks the infection has been more severe, with a significant mortality. The incubation period is about ten days.

(b) Diagnosis, treatment and control

The diagnosis can be made in the laboratory by isolating *C. psittaci* from blood or sputum, though this is rarely attempted. More usually antibodies to the group chlamydial antigen are sought in the patient's serum. The antigen used in the test is present in all chlamydias, so it does not distinguish between the diseases they cause. This does not matter as these diseases have quite different clinical presentations and epidemiologies. The treatment of psittacosis is with tetracycline which, if given early, limits the extent of lung involvement, so may save life.

The health authorities of many countries control or prohibit the importation of wild birds, particularly of the parrot family. They may be subject to quarantine on arrival. Cases of human ornithosis should be followed up epidemiologically to discover their origin so as to impose the necessary controls. Infected birds may be treated or destroyed. Commercial flocks should be bred to be free of chlamydias.

15.2.2 *C. pneumoniae*

The infection due to this chlamydia is similar to that caused by *C. psittaci*. The disease differs from psittacosis in that it depends on person-to-person transmission, with no animal host. When it was first recognized some years ago it was called the 'Taiwan acute respiratory chlamydia', or TWAR.

15.2.3 *C. trachomatis*

Although they are all included within a single species *C. trachomatis* is really a collection of related bacteria that can be distinguished from each other by their antigenic structures and biological properties. The latter include the production of three quite different types of human disease (Box 15.2). Members of the first group cause trachoma, from which the species takes its name. Trachoma is a leading cause of preventable blindness. It is now evident that members of the second group are responsible for a sexually transmitted disease (STD) that is both different from and much more common than lymphogranuloma venereum, an STD described much earlier and caused by members of the third group of chlamydias.

(a) C. trachomatis, *first group*

The chlamydias in this group cause repeated infections of the eyes. The cornea and conjunctiva are involved to produce a keratoconjunctivitis. With the addition of secondary bacterial infections deep scars are produced that lead to blindness. The disease is a major public health problem in parts of Africa, the Middle East, and SE Asia. It is spread by close personal contact mostly within families, and it starts in childhood. Each attack of trachoma is self-limiting but despite the development of antibody there is no immunity. Reinfections lead to irreversible damage and though the chlamydial process usually ceases in adolescence the deformity already produced predisposes to invasion by other microbes. These cause further damage and perhaps loss of sight by the age of 30.

The diagnosis is made by demonstrating inclusion bodies in conjunctival scrapings stained with Giemsa's stain, or by immunofluorescence, or by the growth of the chlamydia in tissue culture. Treatment is with tetracycline, erythromycin or rifampicin, given systemically. Mass campaigns have attempted to treat all the members of a community simultaneously with an antimicrobial applied locally to the eyes. This approach was not successful.

(b) C. trachomatis, *second group*

In some communities these chlamydias are the cause of about half the cases of non-gonococcal urethritis among men and for some cases of acute epididymitis that may complicate it. About 30% of heterosexual men with chlamydial urethritis have few or no symptoms, and the proportion is higher among homosexuals. In women most chlamydial infections are asymptomatic, though the infection may present clinically as a mucopurulent cervicitis or as a painful pelvic inflammatory disease. Infants born of infected mothers may also be infected. Damage to the fallopian tubes can cause infertility or lead to ectopic pregnancies. The clinical spectrum of the infections caused by this group of chlamydias are presented in Table 15.1.

Table 15.1 The range of infections and complications caused by members of the second group of *C. trachomatis* (Box 15.2)

Infections	Males	Females	Infants
Urethritis	+	+	
Proctitis	+	+	
Inclusion Conjunctivitis	+	+	+
Cervicitis		+	
Pneumonia			+

Complications of these infections

in males	in females
Epididymitis	Salpingitis
Prostatitis	Endometritis
Reiter's syndrome	Perihepatitis
Sterility	Reiter's syndrome
	Ectopic pregnancy
	Infertility

+: anatomical types of infections

The diagnosis can be made by growing the chlamydias on suitable cells in tissue culture. These are examined after a few days to detect inclusion bodies. This is a rather expensive and slow way of making a diagnosis, but it is the most sensitive one. Alternatively, smears of material from the site that is involved clinically may be examined for inclusion bodies after it has been stained with Giemsa's stain. This method misses up to half the culture positive cases, so it is very insensitive. A method of intermediate sensitivity uses immunofluorescence to detect the chlamydias in smears of cells taken from the infected part. These microscopic methods do not distinguish between chlamydias that are alive or dead.

Treatment should be given to both sexual partners (and offspring, if indicated) at the same time, to prevent reinfection. Tetracycline is the drug of choice, though erythroymycin is used in pregnancy. Conjunctivitis can be treated with tetracycline drops or ointment. The control of this as with all sexually transmitted diseases depends on responsible sexual behaviour and the early diagnosis and treatment of infections.

(c)   C. trachomatis, *third group*

This group of chlamydias cause an infection called lymphogranuloma venereum (LGV). The disease is nearly always sexually transmitted. LGV begins as a small, painless spot, usually on the penis or vulva. If untreated the local lymph nodes may enlarge and suppurate, and infection gradually spreads into the surrounding tissues where it causes fibrosis. In males inguinal buboes form, and sinuses develop that discharge pus. In females the lymph nodes inside the pelvis may be involved with inflammation round the rectum and vagina. This can lead to stricture of the rectum and the development of a rectovaginal fistula. There is usually some general malaise. If not treated, LGV develops into a chronic disease accompanied by much deformity and disability.

Diagnosis may be by isolation of the chlamydia, or more usually serologically, as described for psittacosis. Treatment is with tetracycline and sulphonamides. The results are good if this starts before there has been much tissue destruction. Control is as for any sexually transmitted disease.

15.3 THE RICKETTSIAS

Rickettsias are small Gram-negative bacteria that multiply inside cells. Only *Rochalimaea quintana* has been grown on cell-free culture media. Nearly all of them are transmitted from vertebrate to vertebrate, by insects. The usual vertebrate hosts are infected and the rickettsias spread into their blood, though they do not seem to suffer as a result. In the circulation they are available to infect other biting insects when they take a blood meal. It is only when the rickettsias reach an unnatural or unaccustomed host, such as a human or louse, that disease results.

Rickettsias are big enough to be seen under the light microscope when stained with Giemsa's stain. They divide by binary fission, and grow in different parts of the host cell. A classification of the rickettsias and a summary epidemiology of the diseases they cause are given in Tables 15.2 and 15.3.

Table 15.2 The groups of infection caused by the rickettsias

The groups and the diseases

A. *The classical typhus group*
1. Epidemic typhus, Brill's disease
2. Endemic (murine) typhus

B. *Tick-borne spotted fevers*
3. Rocky Mountain spotted fever, Mediterranean spotted fever, Kenyan, Indian, Queensland, or Serbian tick typhus, etc.

C. *Mite typhus*
4. Rickettsial pox
5. Scrub typhus

D. *Aberrant forms of typhus*
6. Q (query) fever
7. Trench fever

Table 15.3 An outline of the epidemiology of the infections listed, and as numbered in Table 15.2

Rickettsia spp.	Vector	Natural hosts	Distribution
1. *R. prowazekii*	louse	man	worldwide
2. *R. typhi*	flea	rodents	worldwide
3. *R. rickettsii*	hard ticks[1]	rodents	worldwide
4. *R. akari*	mites[1]	rodents	USA, NW Asia
5. *R. tsutsugamushi*	mites[1]	rodents	SE Asia, Japan
6. *Coxiella burnetii*	airborne[2]	cattle, sheep, goats	worldwide
7. *Rochalimaea quintana*	louse	man	Europe, America

[1] These insects may transmit the rickettsia to their young through their eggs
[2] Q fever is often transmitted between its animal hosts by ticks. Humans are infected by the inhalation of dust that contains the rickettsia

In general, rickettsias are easily killed by heat, desiccation and disinfectants. *R. burnetii*, the cause of Q fever, does not conform. It is particularly resistant to drying, and can withstand pasteurization. Because of this it can survive for many months or years in dry dust. This has important epidemiological consequences.

A curiosity of the rickettsias is that they share antigens with certain strains of Gram-negative bacteria, *Proteus mirabilis* OXK and *Pr. vulgaris* OX2 and OX19. Because these are grown much more safely and easily than rickettsias, they are used as antigens in the Weil-Felix test for the serological diagnosis of typhus. More sensitive and discriminatory tests use antigens prepared from the rickettsias themselves, though these special reagents are not generally available.

With the exception of Q and trench fevers, diseases due to the rickettsias have a common pathology, though there are characteristic variations in the distribution and severity of the lesions. These result from an attack on the lining of blood vessels. Inflammatory cells collect at the site and there may be haemorrhage if the vessels rupture. The 'typhus nodules' so formed are small, but when gathered together in large numbers they can cause, for example, a significant enlargement of the spleen. Small haemorrhages are common in the skin and brain. In tick and mite typhus the site of the insect bite is usually inflamed and may develop into an ulcer with a black base. Peritonitis, pleurisy and pericarditis may complicate scrub typhus, but in this infection the rash is not haemorrhagic.

15.3.1 *R. prowazekii*

This rickettsia is the cause of epidemic typhus. The disease is marked by fever and severe prostration. Epidemics are usually associated with disasters in which large groups of lousy people are herded together in poor conditions (Box 15.3 and Chapter 27). The illness lasts about two weeks, with a rash

---

Box 15.3 The louse at war

Epidemic typhus is common in times of war, and has sometimes determined the outcome. The fall of Prague to the French in 1741 and Napoleon's retreat from Moscow were influenced by it. In World War I 150 000 Serbs and 3 million Russians are said to have died of typhus. It was the cause of many of the deaths in concentration camps in World War II.

---

that appears on the sixth day. Because disaster and deprivation usually go together epidemic typhus is accompanied by a high mortality. The vector louse also suffers and dies from the infection, which is transmitted through its faeces, not by bite. The rickettsias may not be eliminated after a patient has recovered. In later life they can re-emerge to cause a short sharp attack of recrudescent typhus, or Brill's disease. Another disaster may provide the stress that precipitates this. If lice are present at the same time and they feed on an individual with Brill's disease they are infected and can start another outbreak of epidemic typhus. Epidemic typhus will continue to exist if the intervals between successive disasters is less than a human lifespan.

15.3.2 *R. typhi*

Endemic (murine) typhus is similar to though less severe than epidemic typhus. It appears sporadically, and is associated with rat or other rodent infestations in or around particular buildings. The rickettsias are transmitted by rat fleas, and they are excreted in their faeces (as with epidemic typhus) so infection follows contamination of the bite as the flea feeds and defecates simultaneously. The disease is readily distinguished from epidemic typhus epidemiologically, and in the laboratory.

15.3.3 *R. rickettsii* (and other named species)

The spotted fevers are a worldwide collection of diseases that are similar to each other, though with variations in severity. They are causd by a group of related rickettsias that are distinguishable antigenically. The infections are often given names that locate them geographically, such as Rocky Mountain or Mediterranean spotted fevers, Indian or African tick typhus, and so on. As the collective name suggests rashes are a prominent feature. They are spread by hard ticks that normally feed on small wild rodents, more often in rural surroundings. Humans are involved when they invade the territory occupied by the natural hosts of potentially pathogenic rickettsias, and if they are bitten by infected ticks.

15.3.4 *R. akari*

Rickettsial pox is a mild disease that tends to be associated with particular dilapidated buildings. It is caused by the bite of an infected mite. The rash is like that of chickenpox. The rickettsia is maintained by a cycle that involves household rodent pests and the mites themselves.

15.3.5 *R. tsutsugamushi*

Scrub typhus is found inside a triangle with Japan, India and the northern tip of Australia at its apices. In its natural state it is a symptomless infection of mites and small rodents that live in the jungle or in areas of scrub. Larval mites may feed on humans if they invade their territory, and so transmit the infection.

15.3.6 *Coxiella burnetii* (Q fever)

This rickettsia is the cause of a mild disease of cattle, sheep, goats and some wild animals, among which it is spread by ticks. Infected animals tend to secrete the rickettsia in their milk (where it is partially resistant to pasteurization), and they are present in very large numbers in the placenta and fluids at the time of birth. The organism is resistant to drying, so can remain viable as it blows about in the dust when contaminated straw or animal bedding is disturbed. Humans are usually infected by inhalation, and suffer an acute febrile illness with malaise and myalgia. In many cases this develops into an atypical pneumonia, with cough, scanty sputum, and chest pain. Significant X-ray changes are accompanied by minimal physical signs in the lungs. In individuals with abnormal heart valves *C. burnetii* can cause an infective endocarditis. The disease is called Q, for query, fever.

15.3.7 *Rochalimaea quintana*

Trench or shinbone fever is a rare disease that has been recognized in louse-infested armies in the field or prisoners of war who are housed in crowded, unhygienic conditions. There is a debilitating though rarely fatal fever, with pain in the shins. It is spread in the same way as epidemic typhus.

(a)   Diagnosis (of all the rickettsial diseases)

Although rickettsial isolation is the most precise way of making a diagnosis, it is rarely attempted. Accidental laboratory infections are common, and both the American Howard Taylor Ricketts and the Pole Von Prowazek died of typhus due to the microbes that now bear their names. Serological methods are normally used. The Weil-Felix test is widely employed, though it is by no means always positive in cases of true typhus. The test depends on the detection of antibodies that react with one or more of the proteus strains OX19 and OXK. Better tests using rickettsial antigens are available in specialist centres.

(b)   Treatment and control

Rickettsias are sensitive to tetracycline and chloramphenicol, and the severity and duration of the infections are reduced by their use. As most forms of the disease have a significant mortality, this is important. Sulphonamides make the diseases worse.

Consideration of the epidemiology of each rickettsial disease will indicate the control measures most appropriate to it. With epidemic typhus, delousing is important; with murine typhus, rodent control; with tick and mite-borne typhus it is necessary to avoid or clear secondary jungle and vegetation in which infected rodents and their ectoparasites live. If exposure is unavoidable protective clothing treated with an insect repellent is worn. Vaccines have been prepared, but are not often used.

FURTHER READING

See Appendix B.

Viruses and the Infections They Cause

Introduction to the viruses $\boxed{\textbf{16}}$

16.1 VIRAL ANATOMY

The largest viruses (the poxviruses) are about the same size as the smallest bacteria (0.2 μm). The smallest viruses are the poliovirus (0.03 μm) and the parvoviruses (0.02 μm). When measured in nanometres the diameters of viruses range between 20 nm and 200 nm. The poxoviruses can just be seen with a light microscope, but the rest are invisible. Their anatomy has finally been revealed by electron microscopy, chemistry and X-ray analysis. The

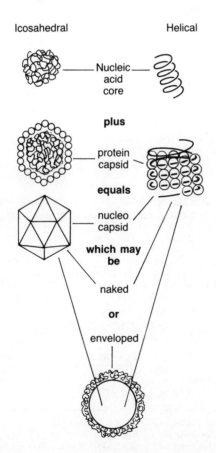

Figure 16.1 The anatomy of viruses with icosahedral and helical symmetries.

detailed anatomy of viruses now provides a satisfactory basis for their classification.

The characteristics used to classify the viruses are the type and arrangement of their nucleic acid, and the nature of the shell into which this is packaged. The genetic blueprint of a virus may be carried on ribonucleic acid (RNA) or deoxyribonucleic acid (DNA). A single virus never has more than one kind of nucleic acid, a feature that separates them from all living things. The molecules of RNA or DNA may be arranged in single or double strands and in a linear or circular form. The nucleic acid core is enclosed within and protected by a layer of protein molecules (the **capsid**) to form a **nucleocapsid** (Figure 16.1). The protein molecules of the capsid are usually arranged over the surface of the core in a symmetrical, geometric fashion. In some cases the capsid is made up of 20 interlocking triangular faces to form an approximate sphere that surrounds the nucleic acid. In geometry this structure is called an icosahedron so a virus of this shape is said to possess **icosahedral symmetry**. Otherwise the arrangement may be spiral (for **helical symmetry**) or the arrangement may be more complex (**complex symmetry**).

A single virus particle or **virion** may consist of no more than its nucleocapsid, so is naked. Alternatively the nucleocapsid is surrounded by an outer shell or envelope. This is usually derived from the membrane of the cell in which the virus was formed. In addition to the **structural proteins** of the capsid there may be other, **functional proteins**. Some of these operate to effect the preliminary attachment of the virus to its target cell, or they assist with this in other ways. With some RNA viruses functional proteins are needed to initiate the process by which copies of this type of nucleic acid are made prior to their insertion into new viruses. Some of these features are depicted in Figure 16.1, and they contribute to the simple classification of viruses given in Table 16.1.

Table 16.1 A classification of the major viruses pathogenic for humans according to the chemical nature of their genetic material, and their outward anatomy

	Genetic material	
Surface anatomy	*RNA*	*DNA*
UN-ENVELOPED (naked)	Enteroviruses Polioviruses Rhinoviruses Rotaviruses	Adenoviruses B19 virus[1] Warts viruses
ENVELOPED	HIV virus[2] Measles virus Mumps virus Rabies virus Rubella virus	Hepatitis B virus Herpes viruses[3] Poxviruses

[1] the virus of 'fifth disease' or erythema infectiosum
[2] the human immunodeficiecy (AIDS) virus
[3] includes the herpes simplex, cytomegalo-, Epstein-Barr (glandular fever) and the varicella-zoster (chickenpox and shingles) viruses

16.2 REPRODUCTION

Viruses reproduce by a process that is quite unique. They lack the equipment to multiply on their own so depend on other living cells to help them. When inside a cell they divert the metabolic activities they find there to the production of new virions. The cells they invade may be those of animals, plants or even other microbes. The viruses that multiply inside bacteria are the **bacteriophages** that play a part in bacterial recombination (Chapter 5).

Other microbes have developed the ability to live and multiply inside foreign cells. The rickettsias and chlamydias also use the metabolic systems of other cells to supplement their own, but they are not so completely dependent on their hosts as are the viruses. An essential difference between viruses and other microbes is that after entry viruses break down into their component parts, so cease to exist as entities. The viral nucleic acid takes control of the cell which is now forced to make new viral nucleic acid, viral protein and other viral components. At some point these parts are assembled into a number, perhaps hundreds, of complete new virions. These may escape suddenly when the cell in which they have been formed breaks open (by **lysis**), or by a process called **budding** if they emerge through the host's cell membrane without rupturing it.

Lysis results in the immediate death of the cell while budding continues for a time before the cell dies. A third outcome of the entry of a virus into a cell is that the viral nucleic acid remains inactive and hides itself somewhere, perhaps in the nucleus of its new host. Such a cell, together with its progeny, is occupied by a hidden or **latent** virus. In response to a stimulus some time (perhaps years) later, the latent viral nucleic acid is activated to produce a burst of new virions, accompanied by cell damage. This is what happens in recurrent herpes simplex. Alternatively the viral nucleic acid may disturb the normal control mechanism that arranges the orderly multiplication of host cells. The result of this may be the beginning of an uncontrolled multiplication that leads to cancer.

Because viruses use the metabolic systems of their hosts to multiply, it is more difficult to design antiviral equivalents of the antibacterial drugs. An attack on a target that might interfere with viral multiplication is also likely to damage uninfected host cells. This is why progress in antiviral therapy has been slow.

16.3 CULTURE AND IDENTIFICATION

Viruses only multiply in living cells. In the laboratory these are provided as whole animals, including the embryos of fertile eggs, or in the form of cells taken from living things that are grown and maintained in glass or plastic containers. The latter process is called **tissue** or **cell culture**. All these methods are expensive and each has disadvantages.

In the case of cell cultures the first disadvantage is the technical difficulty of preparing one culture to support another. When bacteria or fungi contaminate the media used for tissue culture the cultured cells are destroyed before viruses can grow in them. Most clinical specimens that may contain a virus also contain bacteria and perhaps fungi as well. Cell cultures could not be used for routine virology until antibacterial and antifungal drugs began to appear in the 1940s. When these were added to culture media, cell cultures began to displace the other methods. Used in this way antimicrobials also made possible the production of the large quantities of virus needed to prepare vaccines. Poliovaccine was an early example.

Cell cultures may be prepared from almost anything that is alive. Tissues are broken up mechanically and chemically to produce single cells. Sometimes it is possible to make these cells multiply in artifical media to produce an immortal source of cultured cells. For routine use the cells are made to adhere to the inside surfaces of containers where they are covered by a nutrient medium and through the walls of which they can be examined with a microscope. When the cells have settled down in this environment the material to be examined for the presence of a virus is added to the container. This material may be a specimen from a patient thought to be suffering from a virus infection. The culture is then examined daily for evidence of viral multiplication.

If a lytic virus is present the cultured cells are killed progressively as new virus released from the first cells infected invade and destroy those that remain, until none are left. In the case of a virus released by budding damage may be less obvious, but in either case microscopic evidence of cell damage (a **cytopathic effect**) is sought. If cell damage is too slight to be visible directly, other methods are employed to detect it. The process is slow, taking several days to complete. Several more days may then be needed to identify a virus after it has been isolated.

The traditional way to identify a newly isolated virus is by a **virus neutralization test**. When viruses are mixed with an antibody prepared against them they are 'neutralized' because they can no longer enter target cells in which to multiply (Chapter 2). A series of cell cultures is set up, and to each is added one of a series of antibodies prepared against different viruses. A suspension of the virus to be identified is then added to all the cultures, and these are examined daily. The culture in which the virus fails to grow must contain the antibody that matches it, so it is identified.

Before a virus can enter a cell, either in the laboratory or in nature, it must identify that the cell is of the right kind, and adhere to the outside of it. As was the case with phagocytosis and with antigen-antibody reactions (Chapter 2), the surfaces of the virus and the cell must possess complementary molecular structures that fit together like a lock and its key. This requirement for mutual recognition explains why a virus cannot enter any cell it pleases. Dogs do not suffer from measles or humans from distemper because of this, and in humans polioviruses attack the cells of the central nervous system and not of the lung. This specificity is a problem for diagnostic virus laboratories. Even when cultures of several different types of cell are used it is not possible to provide the range necessary to accommodate all pathogenic viruses.

For this reason it is important to have some idea of what group of viruses might be responsible for an infection before an attempt is made to isolate it.

A diagnostic bacteriology laboratory can quite easily employ culture media on which most important bacterial pathogens can grow. In this way bacteriologists trawl the microbiological ocean to see what is there. In contrast virologists stalk their prey individually, with a spear-gun. Bacteriologists may grumble when throat-swabs reach them with the unhelpful request '?bacteria', but something useful may still be done. A virologist is more likely to react to the request '?virus' by rejecting the specimen out of hand.

Virologists must contend with expensive cell cultures that cannot provide for all possible pathogens. Their difficulties are compounded by the fact that the result of a culture is unlikely to be available until after the patient has recovered from an infection, or has died of it. To overcome these problems a variety of rapid methods have evolved, though these increase rather than reduce the need for accurate clinical information about the patient, otherwise the wrong one may be used. The tests may involve the electron microscope (EM) or techniques like fluorescence microscopy that use tracer molecules or probes to seek out microbes. These probe molecules are labelled with something easily detected to show when a tracer probe has found its target, to indicate where this is, and how much there is of it. These tests are described in more detail in Chapter 29.

Many clinical specimens are not suitable for examination by EM because they contain only a few viruses hidden among a log of debris, in insufficient volume to permit the virus to be concentrated before examination. A specimen with a large volume (faeces) or a high concentration of virus and not much else (fluid from a chickenpox vesicle) may be examined usefully. Even when the examination is positive it is usually only possible to say that a virus of a certain group is present (one of the herpes viruses for example), but not to identify it more precisely.

16.4 VIRAL PATHOLOGY

Virus infections present in ways that vary enormously. They may be acute, subacute, subclinical, latent, persistent or recurrent. To produce disease a virus must kill, or interfere with the normal multiplication of cells. Cell death may cause symptoms on its own when, say, nerve cells are destroyed, or symptoms may result from the host's response to the presence of dead or dying tissue. Viruses may enter the body by inhalation, ingestion, inoculation (bites of aimals or insects, needle puncture, injury) or as a result of intimate contact. New routes have been opened up by the practices of blood transfusion and organ or tissue transplantation. The surface of the virus must react with the surface of its target host cell, and attach to it. The presence or absence of the appropriate receptor on a cell determines if that cell is sensitive or resistant to a particular virus. Individual cells may have tens of thousands of these specific receptors on their surfaces.

Viruses often multiply at the point of their entry into the body. The disease that results may be limited to this site, as happens with the viruses that cause diarrhoea and vomiting, or the common cold. Otherwise viruses may spread from the initial site of primary multiplication through the bloodstream or lymphatics to reach distant, secondary targets. In these cases the primary multiplication may be symptomless, and any disease that appears is the result of damage at the secondary site. This is what happens in poliomyelitis. Some viruses, rabies for example, spread along nerves.

As with bacteria, the viruses that infect the human race defy orderly description. Viruses with similar structures may cause widely different diseases, while quite different viruses may produce infections that appear to be the same. Fortunately some simplification is possible. Certain viruses cause disease because they attack the cells of the respiratory tract. For descriptive purposes these are considered together as the respiratory viruses. The same applies to a group of gastrointestinal viruses. Others fall into groups because they cause classical infections of children, because they are spread by insects, or because they attack a particular organ. When the individual groups of infections created in this way have been dealt with those that remain form the rather untidy residue covered in Chapter 23.

FURTHER READING

See Appendix B.

The respiratory viruses 17
(Influenza, Parainfluenza, Respiratory Syncytial, Adeno-, Rhino-, Corona- and Reo- viruses)

17.1 INTRODUCTION

Most of the infections due to respiratory viruses are mild or subclinical. They are also very common. In advanced countries more working time is lost due to 'colds' than to any other illness, and very large sums of money are spent on remedies for them. Infants and the elderly tend to suffer more severely from respiratory infections, and this is particularly true in the less developed parts of the world.

The viruses included in this cateogry and their associated infections are listed in Table 17.1. Any of them (together with some bacteria) may cause simple upper respiratory tract infections (URTIs), or 'colds'. In more severe infections symptoms and signs develop that may give a clue to the virus responsible. The overlap is great, however, and only in an epidemic is it wise to make a firm clinical diagnosis without help from a laboratory. Other microbes may attack the respiratory tract after it has been damaged by a virus, so secondary bacterial infections are common.

Table 17.1 The human respiratory viruses, the number of types of each and the infections they cause if illness progresses beyond the stage of a simple 'cold'

Virus	No. of types	Characteristic infections
Influenza	3	Influenza
Parainfluenza	4	Croup, pneumonia
Respiratory syncytial virus	1	Bronchiolitis, pneumonia
Adenoviruses	41	Pharyngitis, conjunctivitis
Rhinoviruses	>100	Colds
Coronaviruses	3	Colds
Reoviruses	3	Colds, diarrhoea

17.2 THE INFLUENZA VIRUSES

Influenza A infections may be sporadic, but the virus has a tendency to cause epidemics or pandemics. In a sharp epidemic the relatively small proportion of severe or fatal cases are more obvious, so influenza A has a bad reputation. The type B virus tends to cause smaller outbreaks but it may also be lethal, particularly among the elderly. The envelopes of these viruses are equipped with two functional proteins (Chapter 16) that operate to attach them to their target cells. These proteins are antigenic, and after an infection the neutralizing antibodies that develop serve to protect an individual from a second attack by the same virus.

From time to time the functional proteins change their antigenic structures. Major changes (antigenic shift) take place less often, while minor changes (antigenic drift) are more common. Epidemics or pandemics of influenza A happen when new functional proteins appear, to which the human population has no immunity (Box 17.1). A full description of an influenza

---

Box 17.1 Pandemic influenza

In 1918 and 1919 two or three waves of influenza swept round the world. In a few months over 20 million died, more than had been killed in the 1914–18 war. The first wave was of typical influenza, with a low mortality. For most of those attacked in the second and third waves the disease was similar, but between 10% and 20% of them suffered much more severely, and about 5% died of an acute lower respiratory disease with an intense, so-called heliotrope cyanosis. No convincing explanation has emerged to account for these exceptional events.

It is possible that the cause was a major antigenic shift in the influenza A virus between the first and second wave of the pandemic. If this is what happened those infected in the first wave had no neutralizing immunity to the new virus, so might be attacked for a second time. Athough one or both the functional proteins were new, the remainder of the virus was the same. Individuals attacked for a second time already had antibody to the rest of the virus, and this would have been boosted by the repeat infection. The simultaneous presence of large amounts of antibody and antigen provides conditions for antigen-antibody complexes to form in the lungs. These would cause an intense inflammatory response that might account for the events.

---

virus gives its major type (A or B), the location at which it was first isolated and strain number allocated there, the year of isolation, and an indication of the nature of its functional surface antigens. An example is A/Singapore/1/1985(H3N2), where H stands for haemagglutinin and N for neuraminidase, the names of the two functional proteins.

(a) Pathogenesis

Influenza is transmitted by the respiratory route, by inhalation or by direct contact. Infection leads to widespread destruction of the lining of the respiratory tract and the release of a large amount of interferon (Chapter 2). These events account for the symptoms of the disease. The incubation period is typically 1–3 days, and the disease is marked by fever, shivering, headache and aches in the back and limbs accompanied by a cough and chest pain. The acute illness last three or more days, though the patient usually feels tired and run down for much longer. Physical examination reveals a fever with conjunctival and pharyngeal inflammation.

A rare but serious complication is a pneumonia due to the virus itself. This is seen most often in the elderly or debilitated. Secondary bacterial pneumonia is more common when bacteria (for example *H. influenzae*, *Strep. pneumoniae* or *Staph. aureus*) adhere to, and invade respiratory mucosa damaged by the virus. Reye's syndrome (disturbance of the brain and degeneration of the liver) is occasionally seen following infections in children. This condition has been linked to the consumption of aspirin and it may be that an attack of influenza provides a reason for giving the drug.

(b) Diagnosis and treatment

Most cases of influenza are diagnosed clinically and, other than in an epidemic, with considerable imprecision. Laboratory diagnosis is by virus isolation from nasopharyngeal aspirates or throat swabs. Alternatively virus antigen may be detected in respiratory secretions, or a serological approach may be used. Although of no use to the individual victim, some laboratory activity is to be encouraged. The information gained can indicate the emergence of a new variant of the virus, so predict an epidemic and show the need for a change in the vaccine. This information is collected and collated internationally by the World Health Organisation.

In uncomplicated cases symptomatic treatment and rest is all that is required. Hospital admission may be necessary for supportive therapy if respiratory function deteriorates too much, and antimicrobial drugs are used to treat secondary bacterial infections. If given early the antiviral drug amantadine may be useful in the treatment of influenza A only and it may also be used prophylactically (Chapter 30). Vaccines are of some value, but they have to be changed and readministered year by year as antigenic drift and shift alter the functional proteins of the viruses. It is usually given to high-isk groups that include the elderly or those with chronic lung or heart disease.

17.3 THE PARAINFLUENZA VIRUSES

The four types of this virus are responsible for about a quarter of severe non-bacterial respiratory infections in children. Children under three may suffer from croup with a barking cough accompanied by noisy breathing. This indicates obstruction of the airway (Figure 17.1). If the obstruction is severe, assisted ventilation may be necessary. The diagnosis is made by immunofluorescent examination of nasopharyngeal aspirates, or by virus isolation.

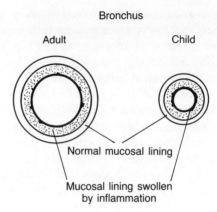

Figure 17.1 Why infants and small children suffer more severely from viral chest infections. The amount of swelling in each bronchus is the same but the airway is much more severely restricted in the very young, in whom it is also more easily blocked by secretions.

17.4 RESPIRATORY SYNCYTIAL VIRUS (RSV)

RSV may cause a mild illness in older children or adults but in infants the infection tends to be more severe. The incubation period is from 1–4 days. In an infant an initial URTI may change into an illness with severe bronchitis and bronchiolitis with respiratory obstruction that may require admission to hospital for respiratory support (Figure 17.1).

Diagnosis is by immunofluorescent examination of nasopharyngeal aspirates. Treatment is supportive, though the inhalation of an aerosol of the antiviral drug ribavirin may be beneficial. The disease can spread inside hospitals. Transmission from one baby to another is more likely to be on the hands of staff than through the air, and mild cases of RSV infection in adults who continue to work may be the source of virus that causes much more severe disease among the children in their care. When an RSV vaccine was tested, those who had been vaccinated suffered more severely from the disease than those who had not.

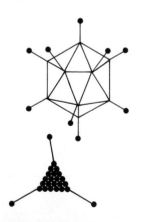

Figure 17.2 A naked icosahedral adenovirus virion, with fibres for attachment to host cells. Each triangular face is made up of 21 protein molecules, with the ones at the edges shared between neighbouring triangles. A fibre is attached to the molecule at each of the 12 'corners'.

17.5 THE ADENOVIRUSES

These naked icosahedral viruses are equipped with fibres that project from the surface of the capsid. These are for attachment to host cells (Figure 17.2). There are over 100 different antigenic forms of these fibres, 41 of which are found in adenoviruses that cause disease in humans. Adenoviruses may cause symptomless infections, mild URTIs or one of the three more characteristic diseases described below.

17.5.1 Acute respiratory disease (ARD)

After an incubation period of 5–6 days individuals suffer a sudden onset of fever, headache, malaise and anorexia. Sore throat and nasal discharge are common. Some who are more ill develop a cough with an X-ray appearance that suggests atypical pneumonia.

17.5.2 Pharyngo-conjunctival fever (PCF)

This infection resembles ARD but with the addition of conjunctivitis of one or both eyes, and enlargement of the local lymph nodes. The conjunctivitis may be severe and resemble the acute haemorrhagic conjunctivitis caused by some of the gastrointestinal viruses.

17.5.3 Epidemic keratoconjunctivitis

In this case a severe conjunctivitis may progress to produce corneal erosions, scarring and opacities. The infection is spread by direct contact between members of families or in schools, or by shared items of equipment such as welders' goggles or on inadequately disinfected ophthalmological instruments.

(a) Diagnosis and treatment

The virus may be identified in clinical specimens by isolation, by electron microscopy or by a search for specific viral antigens. Serological diagnosis is also available. The treatment is symptomatic.

17.6 THE RHINOVIRUSES

There are over 100 different antigenic types of rhinovirus that may infect humans, and others are found in animals. They are an important, but not the only, cause of the 'common cold'. The symptoms of a cold result from the invasion of the cells lining the respiratory tract, initially of the nose. This causes sneezing, and an overproduction of secretion, first watery, then mucoid and finally purulent. The latter is due to the shedding of dead superficial tissue, or it represents the establishment of secondary bacterial infection. The infection may descend into the throat, larynx, trachea and bronchi, with extension into the sinuses and middle ear. Fever is uncommon.

Experiments have shown that draughts or wet clothes do not predispose to colds. The infection is transmitted between individuals rather ineffectively by indirect exposure through the air and more effectively by physical contact, typically involving the hands. Because mild and subclinical infections are so common it is possible to 'catch' a cold from someone who does not appear to have one and some individuals seem to escape when surrounded by them. This gives comfort to those who believe in such prophylactic measures as eating onions, or large doses of vitamin C.

The incubation period is 12 to 72 hours. Immunity is short-lived which is why mild or subclinical attacks are common. Because of the large number of antigenic types, a useful vaccine is unlikely to be developed. Laboratory diagnosis is rarely attempted.

17.7 THE CORONAVIRUSES AND REOVIRUSES

These cause colds rather later in the winter than those due to rhinoviruses, but otherwise they are indistinguishable. The reoviruses may also cause gastroenteritis in children. Laboratory diagnosis is seldom attempted.

FURTHER READING

See Appendix B.

Some viral infections of childhood

18

(Measles, Mumps, Rubella, Fifth Disease)

18.1 INTRODUCTION

Within living memory in even the most advanced countries virtually every child suffered from scarlet fever, measles, mumps, German measles (rubella), fifth disease (erythema infectiosum), chickenpox (varicella) and whooping cough (pertussis). Diphtheria was still common. Taken together these diseases caused an enormous amount of morbidity, much permanent disability and a highly significant mortality. A variety of socio-economic measures and in particular the introduction of effective vaccines have altered the picture completely. Three of the diseases listed are caused by bacteria (Part 3), and chickenpox is caused by a herpes virus (Chapter 20). The other four infections are discussed here.

18.2 MEASLES

The measles virus is a close relative of those that cause mumps and para-influenza but the diseases are quite distinct. Measles is highly infectious, with an incubation period of 10–14 days, or slightly longer in adults (Box 18.1). The disease starts as an upper respiratory tract infection (URTI) with red eyes, running nose, sneezing and coughing. Off-white Koplik's spots 'like

---

Box 18.1 Measles or rubeola

Measles is one of the most infectious of diseases. It is found worldwide, and has been known for at least 2000 years. In 1846 a Danish physician called Panum described an epidemic of the disease on the Faroe Islands. He noted that it was contagious, with an incubation period of about two weeks. He also observed that an attack of measles gave lifelong immunity, as those who had suffered from it the last time it reached the islands 60 to 70 years earlier were not infected again.

---

grains of salt on red velvet' appear on the inside of the cheeks opposite the back teeth. This is where the virus begins to multiply. On the fourth or fifth day of the illness a blotchy red rash appears on the back of the neck and over the next two days this spreads to cover the rest of the body. After another four or five days the rash fades and the patient is no longer infectious.

Serious complications are unusual in well-nourished children, but middle-ear infections, bronchitis and secondary bacterial pneumonia are common among the undernourished and the immunodeficient. Measles encephalitis is a rare complication, and blindness may be caused in children whose stores of vitamin A are inadequate. For these reasons measles is an important contributor to infant and child morbidity and mortality in the less developed parts of the world.

The infection is readily diagnosed at the bedside, though laboratory tests are available, if required. It is prevented by immunization, and measles vaccine is commonly added to those for mumps and rubella to make the MMR vaccine.

18.3 MUMPS

About a quarter of cases of infection with the mumps virus is subclinical. For the other three-quarters an incubation period of 18–21 days is followed by the appearance of tender, enlarged salivary glands, most commonly of one or both of the parotids. This may be accompanied by a mild URTI. The infection can be transmitted to others for about two weeks, for as long as the salivary glands are enlarged. In adolescents and adults the testes and less often the ovaries may be involved. The painful swelling of these organs may lead to sterility. Mumps meningitis is a rare complication. Laboratory tests are available, but are rarely used. The disease is prevented by immunization with the MMR vaccine.

18.4 RUBELLA

Rubella (German measles) is a mild disease whose importance lies in the fact that an infection in the early stages of pregnancy may involve the fetus and cause serious congenital abnormalities. Inapparent subclinical infections are common.

The incubation period lasts between two and three weeks. In young children the appearance of a blotchy pink rash may be the first and only sign of the infection. In older children and adults there may be a period of mild URTI before the rash appears. This fades after 48 hours. Adults may suffer from arthritis of the knees, wrists and hands. Infected individuals may transmit it to others from seven days before the rash appears to five days after, but infants born with a congenital infection are infectious for many months.

Maternal infections in the first 16 weeks of pregnancy may involve the fetus. The result can be fetal death and abortion, and if the infant is born alive it may suffer from blindness, deafness, cardiac abnormalities, or mental retardation. These defects are more frequent after infections in the first four weeks of pregnancy, when about 70% of fetuses that survive the infection are born with abnormalities. Damage to the fetus becomes less common as the pregnancy progresses, and the incidence of congenital defects falls to 5% or less with infections after the 16th week.

Serological tests are available to detect antibodies to the rubella virus that result from natural infections or vaccination. Tests can also indicate if an infection was recent or not. When rubella strikes in pregnancy these tests are of critical importance to give accurate information if termination is an option. Rubella immunization in childhood makes such painful decisions unnecessary. The vaccine should not be given in pregnancy.

18.5 FIFTH DISEASE

This disease (otherwise erythema infectiosum or the human parvovirus infection) was named in the nineteenth century as one of six childhood infections accompanied by a rash. Another name that describes the unusual appearance of the rash is 'slapped cheek syndrome'. The incubation period is very variable (4–20 days), and the disease somewhat resembles rubella. The red cheeks are an obvious diagnostic sign and they may be accompanied by a lacy pink rash on the body that lasts for several weeks, particularly if there is exposure to the sun. Patients are infectious for a few days before the rash appears. Immunosuppressed patients may suffer from chronic infections, and continue to be infectious for a long time. As with rubella, arthritis may develop after the acute stage of the infection.

The microbe responsible for the disease was unknown until quite recently. In 1975 a 'new' virus was found in a blood donation, by electron microscopy. A little later this was discovered to be the cause of fifth disease. It was then noticed that the virus also attacks cells in the bone marrow that are responsible for the production of red blood corpuscles. This passes unnoticed in normal individuals in whom there is plenty of reserve. In people whose blood production is already deficient, in sickle-cell disease or thalassaemia for example, the infection can cause a severe anaemia. Although a fetus infected in pregnancy may not survive, the virus does not seem to cause fetal abnormalities among survivors. It has not yet been grown in the laboratory, so there is no vaccine. A diagnostic laboratory test is available.

FURTHER READING

See Appendix B.

19 The gastrointestinal viruses

(Gastroenteritis, Polio-, Coxsackie-, Echo- and Entero-viruses)

19.1 INTRODUCTION

The gastrointestinal viruses are acquired by ingestion, and they multiply in the cells that line the gastrointestinal canal. They are shed in the faeces, so the route of infection is faecal-oral. The infections they cause fall into two categories. In the first category multiplication of the **gastroenteritis viruses** causes significant damage to the lining of the gut, so diarrhoea and vomiting are the principal symptoms. The second category of infections are due to the **enteroviruses**. These can spread beyond the intestine to produce general (systemic) illnesses. Damage due to multiplication in the gut is less severe, so they cause few if any gastrointestinal symptoms. The viruses that belong to each of these categories are listed in Box 19.1.

Box 19.1 The gastrointestinal viruses

	No. of types[1]
Gastroenteritis viruses	
Rotaviruses	4
Norwalk[2]-like viruses	Several
'Small round structured viruses' and others	Several
Enteroviruses	
Polioviruses	3
Coxsackie[3] A	23
B	6
Echoviruses[4]	32
Enteroviruses types 68–72[5]	5

[1] More await discovery, some have been reclassified.
[2] Discovered in Norwalk, Ohio, USA.
[3] Discovered in Coxsackie, New York State, USA.
[4] Enteric Cytopathic Human Orphan (ECHO) viruses (enteric, from the gut; cytopathic, kill cells in culture; orphan, not at first associated with disease).
[5] Enterovirus type 72 is the hepatitis A virus (Chapter 21).

19.2 THE GASTROENTERITIS VIRUSES

19.2.1 The rotaviruses

The rotaviruses belong to the group of reoviruses some of which also cause mild upper respiratory infections (Chapter 17). Rotaviruses are very important causes of diarrhoea in infants. The nucleic acid core of each virus particle is surrounded by a two-layered capsid. Seen under the electron microscope the complete virus has some resemblance to a wheel (Latin *rota* = a wheel). The virus multiplies in the cells that line the small intestine. When numbers of these cells die and are shed there is interference with the absorption of nutrients and water from the gut. If the area of damage is large enough, diarrhoea and perhaps vomiting is the result. The severity of the infection varies from asymptomatic cases through a slight looseness of the stools to severe diarrhoea accompanied by frequent vomiting. The disease is most severe in children below the age of two, in whom potentially lethal dehydration may be missed (Box 19.2). An oral rehydration mixture (Chapter 11 and Box 11.4) may be lifesaving in these cases. The disease in adults is more often asymptomatic. The incubation period is between 1 and 4 days.

Box 19.2 Death from diarrhoea

At least three million young children born in less developed countries die each year from diarrhoea. Death may be rapid. from dehydration, or come more slowly as the result of malnutrition to which repeated attacks of diarrhoea contribute. About half the cases are due to viruses.

In more advanced countries diarrhoea kills less often, but is second only to the common cold as a cause of lost working time. It has been estimated that it is the cause of about 25 days lost from work or school for every 100 members of the population, every year. A significant proportion of this diarrhoea is caused by viruses.

Diagnosis is by electron microscopy of the faeces, or by the application of one of the rapid 'probe and label' tests described in Chapter 29. Treatment is by rehydration, as required. Outbreaks among children in hospitals often involve the staff who care for them. Prevention is by scrupulous hygiene and the avoidance of overcrowding. It is not easy to achieve hygiene among children or the elderly in whom diarrhoea and vomiting is not easily controlled. Breastfed infants have some protection against infection in infancy.

19.2.2 The other gastroenteritis viruses

The disease associated with the other gastroenteritis viruses listed in Box 19.1 resemble that caused by the rotaviruses, though on average the symptoms are less severe.

19.3 THE ENTEROVIRUSES

The enteroviruses belong the same group as the rhinoviruses (Chapter 17). They differ from them by being stable in acid and resistant to bile, necessary preconditions for microbes that operate in the intestine. First these viruses enter and multiply in the cells that line the gut and in the associated lymphatic tissues. This initial infection is limited and usually symptomless. Enteroviruses may spread from these sites into the bloodstream and multiply secondarily at some distant site, to cause disease. This is unusual however, and most enterovirus infections are asymptomatic.

Enteroviruses from the intestine pass out in the faeces, into which they may be shed for several months. Transmission is by the faecal-oral route and is particularly efficient within families and in communities with poor hygiene. The viruses survive quite well in sewage from which they may spread to contaminate drinking water, or shellfish.

When symptoms do develop they usually amount to no more than a mild fever. Less often this progresses to more obvious disease, with characteristic symptoms and signs. These more severe infections fall into certain patterns that are associated with particular viruses, as indicated in Table 19.1.

Table 19.1 Diseases that may be associated with particular enteroviruses. There is a considerable overlap between them

Disease	Enterovirus
Simple fever	Most
Meningitis	Most
Paralysis	Polioviruses
Herpangina; hand, foot & mouth disease	Coxsackie A
Heart infection, Bornholm disease	Coxsackie B
Haemorrhagic conjunctivitis	Enterovirus 70
Hepatitis A	
(Chapter 21)	Enterovirus 72
Severe neonatal disease	Echoviruses, often type 11

19.4 THE POLIOVIRUSES

When polioviruses spread beyond the primary site of infection in the gut they cause mild feverish illnesses. If they also invade the central nervous system they cause meningitis and may also attack and destroy nerve cells, particularly those that operate muscles. When this happens the form of paralysis that results is called **poliomyelitis**. Loss of function of the respiratory muscles is the most serious and potentially lethal complication. Among those who survive there may be a little improvement for about a year as cells damaged by the inflammatory reaction, but not directly attacked by the virus, recover. Perhaps only 10 of each 100 people infected by a poliovirus experience any illness at all, and only one of them develops paralysis. The remainder have mild or subclinical infections that immunize them against future attacks by the same virus. The incubation period between exposure and the begining of illness is 7–14 days.

In unhygienic societies infection is universal and takes place early in infancy, usually under the protection of maternal antibody. This ensures that virtually all the infections are asymptomatic. When social development produces improved living conditions infection is delayed until after maternal antibody has vanished. Disease then appears among older infants who suffer from 'infantile paralysis' as poliomyelitis was once called. With further social and hygienic development infection is delayed even more and epidemics of poliomyelitis began to appear among older children, adolescents and adults. This progression explains why poliomyelitis was once thought of as a disease of temperate rather than tropical parts of the world.

Poliomyelitis is controlled by immunization. Jonas Salk produced the first poliovaccine in 1954. His killed vaccine contains all three polioviruses and three doses of it are given by injection initially, with boosters from time to time. This vaccine stimulates the production of antibody in the blood so prevents the spread of the virus from the gut to the central nervous system. It does little to prevent the natural infecton in the gut. Albert Sabin's rival vaccine appeared in 1957. This contains living strains of all three viruses that have been modified so that although they can still multiply in the gut they no longer attack the central nervous system. The vaccine is given by mouth and it mimics the natural infection, but with minimal risk of disease. A disadvantage is that, as with natural infections, the vaccine virus can spread from person to person and in doing so may regain some virulence. It is cheap and easy to apply but some national authorities prefer the injection form, particularly for use in tropical countries (Chapter 31). Polioviruses only infect the human race, so effective universal vaccination would eradicate the disease. The World Health Organisation has committed itself to this objective by the year 2000.

19.5 OTHER DISEASES ASSOCIATED WITH ENTEROVIRUSES

Bornholm disease, otherwise called epidemic myalgia or pleuodynia, is caused by the Coxsackie B viruses. It was first observed in 1930 on the Danish island of Bornholm. Sufferers complain of fever, headaches and pains in the muscles of the chest and abdomen that last for about a week. The pain may be so severe that it has been called the 'Devil's grip'. Coxsackie B viruses may also cause infections of the heart.

Some Coxsackie A viruses and echoviruses cause feverish illnesses with a rash. If this is accompanied by painful ulcers in the mouth the condition is called herpangina, and if the rash on the hands and feet develops into tiny blisters the disease becomes hand, foot and mouth disease.

Since 1970 outbreaks of acute haemorrhagic conjunctivitis due to enterovirus type 70 have been reported from SE Asia, India, Japan and Africa. The virus is only found in discharges from the eyes and not in the faeces. Enterovirus type 72 is the cause of hepatitis A. This is discussed with the other hepatitis viruses in Chapter 21.

Neonates may suffer from severe generalized disease due to enteroviruses, particularly echovirus type 11. Outbreaks of this infection have caused deaths in hospital nurseries.

FURTHER READING

See Appendix B.

The herpes viruses | 20

(Herpes simplex, Varicella-zoster, Cytomegalo- and the Epstein-Barr viruses)

20.1 INTRODUCTION

There are many different herpes viruses only some of which cause infections in the human race. The four most important are listed in Box 20.1, together with the diseases they cause. All of these viruses are enveloped and appear identical when seen with an electron microscope. They cause initial

---

Box 20.1 Human herpesviruses and their diseases

Herpesvirus types 1 and 2 (HSV1, HSV2): herpes simplex, primary and recurrent

Varicella-zoster virus (VZV): chickenpox and shingles

Cytomegalovirus (CMV): CMV infection

Epstein-Barr virus (EBV): infectious mononucleosis or glandular fever

---

primary infections of different kinds in which some of the cells attacked are not killed outright and inside which the virus persists in a latent form. Some time later these latent viruses may be activated to cause recurrent **secondary infections**. Individuals with secondary infections can transmit the virus to others who have not yet been infected. Herpes virus infections are very common, and many of them are asymptomatic.

20.2 THE HERPES SIMPLEX VIRUSES (HSV)

Minor antigenic differences separate the herpes simplex viruses types 1 and 2. They do not survive well outside the body, so transmission between individuals is unusually the result of close interpersonal contacts. The donor individual may have a primary herpesvirus infection, or a second asymptomatic or mild recurrent infection. A single individual may be infected with HSV1 and HSV2 at different times, but as this seems to happen infrequently it may be that an infection with one virus gives some protection against the other. Primary infections above the waist are often due to HSV1, and those below it to HSV2, but oro-genital sexual activity blurs this distinction.

Infections are common and in many societies most babies are born with antibody acquired passively from their mothers. Contacts with relations and friends in the first few months of life often lead to primary herpesvirus infections. When this happens in babies that still have maternal antibody the infection is nearly always asymptomatic. If a primary infection is acquired in the absence of maternal antibody it is much more likely to result in clinical disease. Antibody may be absent because there has been a delay of more than six months before a child is exposed to infection, when maternal antibody has disappeared. Alternatively and increasingly in more developed societies the mother may have no antibody because she has never been infected with the virus.

Primary infections vary with the part of the body involved but they are all marked by a painful rash with small blisters or vesicles that turn into shallow ulcers. Typical sites for these are round the lips (particularly in children as a result of kisses), the eyes and the genitalia. Previously uninfected health care workers may acquire primary infections from patients who themselves suffer from primary, or from inconspicuous secondary herpes. In such cases an unusual but painful infection called a herpetic whitlow involves the fingers of doctors, dentists or nurses who have not worn gloves when exposed to oral or genital secretions.

An infant may be born to a mother who has acquired a primary genital herpes infection late in pregnancy. If antibodies have not had time to develop and reach her offspring before birth the infant may suffer a severe neonatal infection. This appears a few days after birth and develops into a generalized form of herpes. This is a rare but often fatal condition. Herpes encephalitis is another rare complication, usually of a primary infection. Before treatment was available this was also nearly always fatal.

At the time of the primary disease the virus travels to the part of the central nervous system that is connected anatomically to the site of infection, where it persists for life. From time to time it may be reactivated in response to such stimuli as fever, sunlight, cold, menstruation, or some other stress. The result is an attack of recurrent, secondary herpes that usually appears at the same site as the original primary infection.

'Cold sores' are the most common kind of secondary herpes. These small vesicular eruptions often involve the skin round the mouth and nose. Pain is usually felt at the site before the lesion develops and the sores themselves are more painful than might be expected from their appearance. They also heal rather slowly. Recurrent herpes behaves similarly at other sites.

Recurrent infections of the eye tend to involve the cornea. The keratitis that develops can be seen when it is stained with fluorescein, and its characteristic branching appearance has given it the name dendritic ulcer. Frequent recurrences may lead to permanent scarring and visual impairment.

In the immunosuppressed, particularly in those whose cellular immunity is deficient, recurrent herpes is more severe. In these circumstances a cold sore may grow to involve much of the face, mouth and the oesophagus. This can happen in patients with AIDS.

(a) Diagnosis, treatment and epidemiology

The diagnosis is usually made clinically. Both the herpes viruses are easily grown in the laboratory from specimens taken from lesions, but special tests are required to distinguish between HSV1 and HSV2.

The antiviral drug acyclovir slows down viral multiplication without a significant effect on host cells. It may be used topically, or given orally or intravenously. It reduces the severity of primary and secondary disease, speeds healing, and limits the amount of virus shed to infect others. It does not prevent the development of recurrent disease. Local applications of another antiviral drug, idoxuridine, have been used in eye infections and herpetic whitlows.

The epidemiology of herpes infections is changing. Improved socio-economic conditions increase the living space available to each individual and limit family size. Intimate contacts between adults and young children become less frequent, so 'safe' infections under the cover of maternal antibody are less common. More people now reach school age or young adulthood with no immunity to the herpes viruses. This has played a part in the emergence of the 'new' problem of genital herpes, and the publicity that surrounds it.

20.3 THE VARICELLA-ZOSTER VIRUS (VZV)

Varicella (chickenpox) and herpes zoster (singles) are the primary and recurrent secondary forms of an infection with VZV. Chickenpox is a highly infectious disease. It is so common in childhood that it is rare in adults, in whom it is more severe. After an incubation period characteristically between 13 and 17 days there may be an initial ('prodromal') illness with fever, sore throat, headache and vomiting. After three days, or as the first indication of infection in younger children, the characteristic rash appears. Blotches appear on the trunk and then on the head and limbs and within hours these develop into vesicles, which look as if they are on rather than in the skin. Fresh crops of blotches and vesicles appear over the next few days. The vesicles rupture and crust over. The crusts taken about a week to separate, and scarring is unusual unless there has been heavy secondary infection. The degree of general illness varies with the density of the rash, which may be itchy.

Complications are rare, but the central nervous system may be involved, usually mildly. Chickenpox pneumonia is more serious and more common in adults. Chickenpox in the immunosuppressed (children with leukaemia, for example) may be very severe, with a dense rash and bleeding. In early pregnancy the foetus is infected and the infant may be born with a deformity. In very late pregnancy the outcome depends on whether the maternal infection pre-dates delivery by more than five days. If less than five days there is no time for maternal antibody to cross the placenta to protect the infant, which may then suffer a severe neonatal infection.

Transmission is from cases of varicella or of herpes zoster, either by direct contact or by the respiratory route. Patients with chickenpox are most infectious just before the rash appears and they remain so until new crops of spots cease to emerge and they are dry and crusted. Individuals exposed to infection who are on steroids or who are immunosuppressed may be given zoster immune globulin (ZIG). This is concentrated from the serum of patients who have recently recovered from herpes zoster, so it contains abundant antibody to VZV. This is injected as soon as possible after contact to prevent or at least reduce the severity of the infection.

Shingles is the secondary form of VZV infection. After an attack of chickenpox the virus persists in parts of the nervous system. If immunity declines at any time in later life the latent virus may be reactivated. Pain and a vesicular rash appear in areas of the skin that correspond to the parts of the nervous system that is the source of the virus. The pain may be severe, and last for a long time after the rash has disappeared. If the eyes are involved vision may be affected.

Shingles is quite common in older people, particularly when they are ill. A number of cases of it may appear simultaneously or over a short period among elderly individuals in homes or hospitals. This may simulate an 'outbreak', and suggest that exposure to shingles causes shingles. This is not so. A person who is not immune to the VZV may acquire chickenpox if exposed to shingles but nobody catches shingles from another case of it.

(a)   Diagnoses and treatment

The virus grows very slowly in tissue culture, but it can be seen in fluid from vesicles examined under the electron microscope. This is useful to establish if a rash is due to one of the herpes viruses in cases in which treatment is intended. Acyclovir is again the drug of choice. If it is used early in cases of shingles the disability caused by pain is reduced.

20.4   THE CYTOMEGALOVIRUS (CMV)

Like HSV, CMV infections are generally acquired in infancy, and are usually asymptomatic. The virus is shed in all body fluids for a long time after a primary infection. CMV is important because it may cause congenital disease and serious infections in immunosuppressed patients. The virus also resembles HSV epidemiologically. In conditions of high socio-economic development more people avoid a CNV infection until they reach adult life. This becomes important when transplant operations are done.

In pregnancy a maternal infection is likely to spread to the fetus, particularly if the disease is primary. Fetal involvement is possible at any stage, and the maternal infections are usually asymptomatic. Most fetuses also suffer an asymptomatic infection, but such infants continue to excrete the virus in their urine for many months after birth, so can infect others. About 5% of them are born with more serious CMV infections. Infants who survive may be deaf and mentally handicapped.

Infections after birth rarely cause symptoms but if there is disease it ranges from a mild fever to hepatitis or an infection that resembles glandular fever. Transmission by blood transfusion or transplant is a problem for patients who have not previously been infected with CMV. Patients who have received a transplant are at special risk of serious infections because they are deliberately immunusuppressed to prevent the rejection of their new organs. The risk of infection can be reduced by using blood that has specially prepared so that it can be stored frozen, or blood from which the white cells have been removed. The first of these procedures inactivates the virus, the other removes most of it. Alternatively blood is taken from donors who have been tested and are negative for CMV. In transplants both parties may be tested to avoid the transfer of an organ from a CMV positive donor to a negative recipient.

(a) Diagnosis and treatment

The virus may be grown in tissue culture from appropriate specimens, or the diagnosis is made serologically. Acyclovir is of little use in CMV infections but a related drug, ganciclovir, is more effective.

20.5 THE EPSTEIN-BARR VIRUS (EBV)

This virus was first isolated from tissue taken from patients suffering from Burkitt's lymphoma. Burkitt's lymphoma is a malignancy that was first described in East Africa (Box 20.2).

Box 20.2 Burkitt's lymphoma

In 1958 D. Burkitt, a surgeon working in East Africa, noticed a series of unusual tumours of lymphatic tissues. These cancerous growths were called 'Burkitt's lymphoma'. Two virologists, M.A. Epstein and Y.M. Barr made a tissue culture from the tumour, and the electron microscope showed that it contained what turned out to be a new herpes virus. It soon became clear that what they had found in this serendipitous way was the cause of glandular fever. It was named the 'Epstein-Barr virus'.

The epidemiology of the infection due to EBV is similar to that of HSV. EBV infection is widespread, and most adults have antibody to it. Infection in infancy and childhood is usually asymptomatic but if delayed until adolescence or adult life it is more likely to be accompanied by illness. The disease that develops is called 'infectious mononucleosis' or 'glandular fever'. People who live in better social conditions are exposed to infection later in life, so glandular fever is more common among those with a high standard of living. In such people the close contacts of early sexual activity are often responsible for the transmission of the virus, so the infection has been called the 'kissing disease' (Box 20.3).

Box 20.3 The kissing disease

Glandular fever emerged as a 'new' disease about the beginning of the twentieth century. People who escaped infection with the Epstein-Barr virus earlier in life began to suffer from glandular fever in adolescence. The abnormal cells in their blood sometimes led to the mistaken diagnosis of leukaemia, which then unaccountably got better. In 1921 the disease was properly identified for the first time, and in 1932 the diagnosis was put on a firm footing when J.R. Paul and W. Bunnell introduced the Paul-Bunnell test.

Glandular fever has an incubation period of 30–50 days. There follows a low-grade fever with tiredness and loss of appetite. The lymph glands are enlarged and the liver may be involved, but severe hepatitis with jaundice is unusual. In 'anginose' glandular fever the throat is very swollen and sore, and a membrane may develop. A rash is noticed in about 5% of patients but if ampicillin is given the proportion rises to nearly 100%. This reaction should not be recorded as allergy to penicillin. Weakness and tiredness may persist for several months, but for most of those attacked serious complications are rare.

There is good evidence that EBV is at least in part responsible for two forms of cancer. Burkitt's lymphoma is found chiefly in Africa and Papua New Guinea. Nasopharyngeal carcinoma is seen in the Orient, or among orientals who have settled in other parts of the world. The association of these malignancies with particular racial groups or geographical locations suggests that other factors combine with EBV to cause them.

(a) Diagnoses and treatment

The diagnosis of infectious mononucleosis is often made clinically. In the laboratory a lymphocytosis is found and at least 10% of the cells are large and abnormal (the 'mononucleosis' of its name, Box 20.3). In 90% of patients a 'heterophile' antibody develops that clumps sheep red blood cells. This is detected by the Paul-Bunnell test. More specific tests use the EBV itself as an antigen. The similar illness caused by the CMV is always Paul-Bunnell negative. Treatment is symptomatic and the use of ampicillin is to be avoided, though if it is used the rash that develops is highly diagnostic. Metronidazole may help to reduce the secondary anaerobic infection that contributes to the severe sore throat in the anginose form of the disease.

FURTHER READING

See Appendix B.

Hepatitis, the human immunodeficiency virus and AIDS

21

21.1 INTRODUCTION

Hepatitis and the acquired immunodeficiency syndrome (AIDS) are discussed together because they are of special interest and concern to health care workers (HCWs). When individuals with some types of hepatitis and those who are infected with the human immunodeficiency virus (HIV) fall ill they may transmit these infections to the people who care for them. The hepatitis viruses are highly infectious. The diseases they cause are sometimes severe and may be progressive. The HIV is not so easily transmitted but at present most or all of those who are infected with it will, in time, die of it.

21.2 HEPATITIS

Many different microbes may cause hepatitis as a part of a more general illness, but some viruses attack the liver as their primary target. These hepatitis viruses cause infections that are individually indistinguishable, but when a number of cases are collected together characteristic differences emerge. So far as the intensity of the acute infections are concerned the differences are minor, but they become more important when the severity of their complications are compared.

In hepatitis the liver is inflamed. The amount of damage done as a result is very variable. The disease may be subclinical or there may be a mild illness with nausea and loss of appetite, a more severe illness with jaundice, or at its most severe, complete failure of the liver. The liver can no longer perform its function and toxic materials begin to accumulate when half of its cells have been destroyed. After an attack of hepatitis liver cells multiply to replace those that have died. The repair process may be accompanied by fibrosis and disorganization, to produce cirrhosis of the liver. Repair is even more disordered if the infection persists in a chronic form, and in these cases some cells may become malignant to produce a primary carcinoma of the liver.

Before modern virological techniques were available two types of hepatitis were recognized. They were called 'infectious' and 'serum' hepatitis, respectively. Now a larger number of hepatitis viruses and the infections they cause are labelled with the letters of the alphabet. Infectious hepatitis is hepatitis A, and serum hepatitis, hepatitis B. As tests for these became available it became clear that some cases of hepatitis were neither A nor B, and initially these were labelled 'non-A, non-B' hepatitis. Further work revealed that this was not a single entity and hepatitis C, D, E and F have now been defined, and a G has been suggested. Some of the characteristics of hepatitis types A–F, so far as they are yet known, are described in Table 21.1. Other hepatitis viruses may still await discovery. Although they all share the liver as their prime target they belong to different groups of viruses. For example, type A virus is an enterovirus and type C belongs to the same group as some of the arboviruses.

Table 21.1 An outline of the diseases caused by the hepatitis viruses

Hepatitis	1	2	3	4
A	2–4	F–O	+	–
B	5–14	B	+ +	+
C	6–9	B	+	+
D	2–10	B	?	+
E	2–9	F–O	+	–
F	6–8	B	?	?

1. incubation periods: weeks.
2. transmission: F–O, faecal/oral; B, by blood
3. severity of hepatitis: + = mild; + + = more severe; ? see text
4. persistence of infection: – no; + yes

The hepatitis viruses have two principal modes of transmission, faecal-oral, and blood-borne. Viruses of the first kind are shed in the faeces, while those of the second are found in the blood, and in other body fluids that contain traces of it. The route of faecal-oral transmission commonly involves a connection between dirty hands or sewage, and food and drink. The transmission of the blood-borne viruses involves more or less intimate contact.

21.2.1 Hepatitis A

Hepatitis A (infectious hepatitis) has a relatively short incubation period (Table 21.1). The disease is typically asymptomatic or it may be passed off as a mild 'tummy upset'. Cases of hepatitis with jaundice are less common, and severe disease is unusual. Trillions of virus particles are shed in the faeces for some days before and for a few days after illness commences, if it ever does. Patients with hepatitis A who have beconme jaundiced are no longer infectious and do not need to be isolated. As expected of a disease with a faecal-oral route of transmission it is more common in less hygienic communities, in which nearly everyone has been infected by the time they reach adult life. In the more developed countries today fewer than half of those who reach adult life have had hepatitis A. When any who have not yet been infected travel to countries where hygiene is less good they may acquire the disease. This is important because the infection is more severe in adults than it is in children.

Travellers may be protected by an injection of immunoglobulin prepared from the serum of donors whose blood contains antibody to the virus. The protection only lasts for a few months. There is now an effective vaccine that gives more solid, long-lasting protection. A serological test is available for the diagnosis of cases of hepatitis A, and for the identification of individuals who have had an infection in the past, so are immune and do not need to be vaccinated.

21.2.2 Hepatitis B

Cases of hepatitis B resemble hepatitis A, but there is a longer incubation period and the acute phase tends to be more severe. The infection is often acquired later in life. The virus is transmitted by the blood-borne route, hence the old name, 'serum hepatitis'. After a clinical or sub-clinical attack about 90% of people develop antibody to the virus, and eliminate it. In the other 10% the virus continues to multiply in the liver. This multiplication involves the production of large amounts of the surface components of the virus together with smaller and variable amounts of whole infectious virus. These individuals are chronic carriers who may transmit the infection to others. The risk of this varies according to the amount of infectious virus present in their blood. A series of tests have been devised to detect 'markers' that indicate the level of this risk (Box 21.1). Chronic carriers may suffer progressive disease and develop the complications of cirrhosis and carcinoma of the liver.

Any body fluid from a high-risk carrier is likely to contain the virus (Box 21.1). In practice only blood, saliva transmitted by bite, semen and vaginal

---

Box 21.1 Markers of hepatitis B infection

HBsAg
A component of the surface of the virus. Its presence in blood after an acute infection indicates the carrier state. HBsAg is not infectious on its own.

HBeAg
Part of the core of the virus. Its presence indicates a high risk carrier, who can easily infect others.

HBcAg
This is the core of the virus. Its presence has the same significance as HBeAg.

HBsAb
Presence indicates immunity to hepatitis B. It is not found in carriers.

(HB: hepatitis B
 Ag: antigen, a component of the virus particle
 Ab: antibody to an antigen produced in an infection or after vaccination)

fluid (all of which contain at least small amounts of blood) have been shown to cause infection. A tiny smear of blood may contain millions of infectious doses, so infection can be transmitted by an invisible trace, if it is delivered in the right place. The virus cannot penetrate intact skin or the linings of internal body surfaces. For this reason it is only spread between individuals who combine intimacy with injury, though the latter may be so trivial as to be invisible to the naked eye. Among those at special risk are the family contacts of carriers, who share such things as toothbrushes or razors, and their sexual partners. Transmission is particularly common between male homosexuals, and intravenous drug abusers who share syringes and needles. Blood-sucking insects provide a possible route of transmission, but investigation has shown that the danger is theoretical rather than real.

Another route of transmission is from a carrier mother to her infant. Infants infected in this way are much more likely to become lifelong chronic carriers of the virus, and to develop the progressive form of the disease. Other routes of transmission involve the unsterile needles of ear-piercers, tatooists or acupuncturists. Medical, surgical or dental procedures offer a uniquely effective route for the transmission of blood, body fluids or tissues from one person to another. The transfer may be from patient to health care worker, health care worker to patient, or from one patient to another through some shared diagnostic or therapeutic tool. It may happen as part of a deliberate intervention or as an accident. The infection has often been transmitted as the result of an accidental stab-wound from used hypodermic needles. This important subject is mentioned again in Chapter 32.

The frequency of chronic carriage varies widely. Among the general public in much of western Europe and the USA rates are usually well under 1%, though it is higher among groups who indulge in high-risk behaviour or who have been subjected to particular kinds of treatment

---

Box 21.2 An improbable blood-brotherhood

In 1963 an Australian and a New York haemophiliac were united as blood brothers in a research institute in Philadelphia. The researchers were looking for clues to the mysterious racial differences in the distribution of cancer. They mixed blood samples that had been collected in many parts of the world.

Something in the blood of the Australian aborigine reacted with that from the New Yorker. This was at first labelled 'Australia antigen' but it was later identified as HBsAg (Box 21.1). The haemophiliac had received multiple blood transfusions. As a result he had developed antibody (HBsAb) to the hepatitis B virus because the donor of one or more of those transfusions had been a carrier of hepatitis B, as was the aborigine.

This chance discovery was a vital step in the elucidation of the cause of hepatitis B.

(Box 21.2). In parts of Africa and Asia the rate of chronic carriage may reach 20%. Hundreds of millions of people in these parts of the world are at risk of progressive disease, and because of the high rate of carriage among young women, large numbers of carrier babies are added to the total every year.

The diagnosis of hepatitis B is made serologically. Blood donors and the donors of organs are tested before their blood is used for transfusion or their organs are transplanted. An immunoglobulin rich in antibodies to hepatitis B is available for the protection of those at risk of infection, particularly after an accident. The availability of a vaccine has made an enormous difference. It can protect HCWs from infection, and the vaccine and immunoglobulin are used together to protect infants born of mothers who are carriers.

21.2.3 Hepatitis C

The hepatitis C virus is the cause of most cases of non-A, non-B hepatitis among patients who have received blood transfusions. The disease and its mode of spread is similar to hepatitis B, although sexual intercourse may be less important. It is found more commonly in the Far East.

21.2.4 Hepatitis D

This is an unusual incomplete virus that can only multiply in cells already occupied by the hepatitis B virus. If an individual is infected with the two viruses simultaneously the double infection is likely to be more severe. Alternatively someone who is already a chronic carrier of hepatitis B may acquire the hepatitis D virus as an added infection. When this happens the patient's condition can deteriorate very rapidly. As the hepatitis D virus is utterly dependent for its existence on a hepatitis B infection, hepatitis B vaccine protects against both viruses.

21.2.5 Hepatitis E

There have been large outbreaks of hepatitis in India and the Far East that were spread by the faecal–oral route, particularly through water. The disease resembled hepatitis A but the type A virus was not the cause. It was found that a new virus, labelled hepatitis E, was responsible. The disease is usually mild, but in pregnancy it is the cause of severe disease with a high mortality.

21.2.6 Hepatitis F

When tests for the hepatitis B and C viruses were used to diagnose cases of hepatitis after blood transfusions there were still some that did not give positive tests for either. This led to the recognition of the hepatitis F virus. Infections caused by it are more common in the Far East.

21.3 THE HUMAN IMMUNODEFICIENCY VIRUS AND AIDS

In 1983 the cause of the acquired immunodeficiency syndrome (AIDS, Box 21.3) was identified as a virus, now called the human immunodeficiency virus (HIV). It belongs to a family of viruses that infect animals as well as humans. Other human pathogens in the group cause certain kinds of leukaemia.

Box 21.3 AIDS

In 1981 the Centers for Disease Control, Atlanta, USA (CDC) noticed an increase in requests for pentamidine. This drug is used to treat a rare infection due to *Pneumocystis carinii*. Investigation revealed a group of male homosexuals who, though previously well, were now unaccountably suffering from unusual infections caused by a number of normally non- or weakly pathogenic microbes, or from a rare malignancy, Kaposi's sarcoma. Among members of this group a common infection was pneumonia due to *P. carinii*. It was clear that something had interfered with these individuals' immunity, and the condition was called the 'acquired immunodeficiency syndrome', or AIDS for short. This abbreviation has now passed into common use as a new word.

21.3.1 The human immunodeficiency virus (HIV)

(a) Pathogenesis

There are two HIV viruses, HIV1 and HIV2. They are distinguished by minor antigenic differences, but they produce the same disease with the same epidemiology, though HIV2 may progress less rapidly. The HIV is blood-borne so it is spread in exactly the same ways as the hepatitis B virus (this chapter, above), with one vital difference. The blood of an individual infected with HIV contains perhaps 100 000 fewer infectious doses in each unit of volume than is the case with hepatitis B. The amount of blood (or other fluid that contains it) that must pass from an HIV-positive person to infect someone else is therefore significantly greater, so the infection is transmitted much less easily.

Inside the body the virus attaches to and enters cells that possess specific receptors for it (Box 21.4). Once inside it may reproduce itself and destroy the cell, or become latent, to be activated later. The cells attacked are

---

Box 21.4 HIV and the CD4 antigen

The receptor on human cells for HIV is an antigen called CD4. This antigen is found on the surfaces of certain cells only, and the HIV can only attack these. The cells concerned are some 'T' lymphocytes (Chapter 2), plus monocytes and some other cells found in the lympathic system and in the brain. As an HIV infection progresses, the number of CD4 lymphocytes in the circulation (most of them also classified as T-helper cells) falls. Counts of these cells made at intervals allows the condition of patients to be monitored. A normal count of CD4 lymphocytes is about $1000/mm^3$. In AIDS counts below 100 are common.

---

certain types of lymphocyte, phagocytes and other cells that belong to the host's immune system. The disorganization or destruction of these cells prevents the development of an effective immune response. In consequence the virus persists, and infected individals continue to carry the HIV and are infectious for the rest of their lives. The attack on the immune system is progressive, but the rate at which it is destroyed varies from person to person.

At some stage in this downward progress the immune competence of infected individuals is reduced to the point that microbes ordinarily of little or no importance begin to cause infections. These microbes may already exist in latent form inside the body of an individual. This is the case with members of the herpes group of viruses, *Toxoplasma gondii* and sometimes the tubercle bacillus. Otherwise they may be part of the individual's normal flora, as with *Candida albicans*, or they can come from someone else as can also happen with the tubercle bacillus, or the pneumococcus. Alternatively, like *Cryptococcus neoformans*, they may originate in the environment.

When these opportunistic infections are treated the individuals' immune systems fail to provide the assistance antimicrobial drugs need if they are to be fully effective. In these circumstances therapy may be only partially successful, or it may fail altogether. Alternatively, or in addition, a variety of malignancies may develop that are normally kept at bay by an intact immune system. The appearance of opportunistic infections or HIV-related malignancies is the beginning of the end as it marks the transition from a symptomless HIV infection to the clinical illness of AIDS. For the convenience of epidemiologists the diagnosis of AIDS is made according to certain strict rules, so the point of transition is subject to precise definition. Individuals may develop illnesses related to their positive HIV status that do not meet the criteria that have been set. They are then said to suffer from the 'AIDS-related complex' of diseases until something else happens that allows a formal diagnosis of AIDS to be made.

Primary HIV infections are mild or asymptomatic. A few weeks after exposure a newly infected individual may experience an illness with fever, enlarged lymph glands and a rash. Because the immune system is disordered it may take several months for antibodies to appear, so a person may be infectious before a serological diagnosis is possible. He or she then appears perfectly well until AIDS develops after a period that is usually measured in years. Among those infected with the HIV a quarter have developed AIDS in the first five years, half in ten, and most of the remainder in the next five to ten years. At this stage in the development of the pandemic and in the absence of a cure it is not known what proportion of HIV-infected individuals who do not die of something else first will escape AIDS, but the figure is likely to be small.

The progression of the infection is more rapid in infants and children, in whom the diagnosis is also more difficult. It seems that about 20% of the infants of infected mothers acquire an infection before birth. More are infected during and shortly after birth to bring the total of the infants born of infected mothers who develop infections themselves to about one-third. The HIV is present in breast milk so paediatricians and others who have for years taught that breastfeeding brings great benefit find themselves in a dilemma. The current view is that, for poorly developed countries only, less harm is done by continuing to breastfeed that by stopping it.

(b) Diagnosis

Other than in special circumstances HIV infections are diagnosed serologically. Serological tests of any sort produce some false positive (and false negative) results. Because of the critical importance of a positive result an HIV infection should always be confirmed by a second test, using a different method. The blood of an infant born of an infected mother contains antibodies that have been transferred from her, through the placenta. These maternal antibodies may be detected in the blood of the infant for up to a year. The serological tests available at present cannot distinguish between this antibody and antibody developed by the infant itself, in response to an infection. This is why, for the first year of life, it is difficult to determine if an infant who tests positive has really been infected or not.

(c) Epidemiology

HIV infection and AIDS are highly emotive subjects. It is virtually impossible to say anything about the transmission of the virus that is not seized upon by some special group as being sexist, racist, or emotionally unacceptable. Pressure comes from those who see HIV infection as a threat to their lifestyles and from the not inconsiderable number who make their living out of it. This group includes armies of researchers, health educationalists and workers for charities together with the manufacturers of the kits and drugs used to diagnose and treat the disease. There is even a group who deny the existence of a link between HIV infection and AIDS.

The HIV seems to have appeared for the first time at some point in the mid-twentieth century. Because similar viruses exist in animals it has been proposed that the human virus was modified in some way from one originally present in monkeys. No doubt if this account were written by a monkey the suggestion would be the other way round! There is evidence to suggest that the human virus first appeared in East Africa, where the epidemic of HIV infection and AIDS is much more advanced than in other parts of the world. HIV2 may have had its origin in West Africa.

There is no doubt that the route of transmission of HIV is overwhelmingly by sexual intercourse. In Africa the two sexes are infected equally, and transmission from an HIV-infected mother to her infant happens in a quarter to a third of such births. When it spread to Euorpe and America the infection established itself among male homosexuals, who are still the group most heavily involved. Infection is also common among intravenous drug abusers of both sexes who share syringes and needles. Spread by heterosexual intercourse is less common. The pandemic has now reached the Far East.

The major factor in the sexual transmission of HIV is the frequency of intercourse with different partners. Because the amount of virus present in body fluids is small another factor may be the existence of abrasions or ulcers on the penis or vulva, or in the vagina or anal canal. These provide direct routes by which HIV can reach unprotected tissues that contain its target cells.

As with hepatitis B, medical, surgical and dental practice provide additional routes for the transmission of HIV. Before tests were available infections regularly followed the transfusion of blood or the transplantation of organs from infected donors. Products made from pooled blood donations were particularly dangerous. A single donation from an infected person contaminated whole batches of such medications as the 'factor eight' used to treat patients with haemophilia. A significant proportion of the haemophiliacs in the Western world were infected in this way, and several thousand of them have already developed AIDS. The introduction of quick and accurate tests for HIV antibody has almost ended these disasters, but the existence of a period before infected people develop the diagnostic antibody remains a cause for concern.

HCWs are often worried by the possibility that they may acquire the HIV by accidental exposure to blood or other body fluids from patients infected with it (see also Chapter 32). Atlhough a risk clearly exists exhaustive studies have shown that it is very small. The reason for this is that a rather large volume of fluid from an infected person must be introduced beneath the skin of an HCW if the infection is to be transferred.

HIV infections among HCWs as a result of their occupation have been documented only after exposure to blood. Although other body fluids contain the virus the amount seems to be insufficient to transmit the infection. Infections have followed exceptionally heavy exposures involving contact between blood and damaged skin, the mouth or the eyes. The number of confirmed cases due to these superficial exposures is too small to make an estimate of the risk. Rather more HCWs have been infected as a result of needlesticks or other penetrating injuries. Records have been kept of incidents in which the sharp objects concerned were contaminated with the blood of patients who at the time of the accident were known to be infected with HIV. In 2629 such incidents 10 individual HCWs (0.38%) acquired HIV infections. This amounts to one infection for every 263 penetrating injuries that involved HIV-positive blood. Many of the accidents recorded were avoidable (Chapter 32).

Most of the infections followed punctures of the skin by hollow (hypodermic) rather than solid (suture) needles. This reflects the greater volume of blood transferred by a hollow needle filled with blood, compared with the smear of blood on the outside of a solid needle that may already have passed through a glove or clothing to reach the skin. Transmission of HIV from infected HCWs to patients is even more rare. These low frequencies may be compared with the outcome of similar accidents that have involved patients or HCWs who are carriers of hepatitis B. Surveys have shown that about a quarter of unvaccinated HCWs acquire hepatitis B when they are accidentally inoculated with blood from carriers of the virus. A significant number of patients have also been infected by HCWs who were themselves carriers of the hepatitis B virus (Chapter 32).

(d) Treatment and control

There is no specific treatment for HIV infection, and no vaccine is yet available. The virus changes its antigenic structure rather easily so a vaccine may be a long time coming. The drug azidothymidine (AZT) is beneficial as it reduces the incidence of opportunistic infections in AIDS and so prolongs life. At one time it was thought that it might also delay the transition from an HIV infection to AIDS, but this is now in doubt. Other drugs are being developed, and work on vaccines continues.

21.3.2 AIDS

AIDS is the end stage of an infection with the HIV. A strict definition of what constitutes AIDS is imposed because accurate figures for the incidence of the disease are required for a number of purposes. It is unusual for patients with AIDS to survive for more than five years after the diagnosis is made.

AIDS is the result of the collapse of parts of the immune system. It is associated with a bewildering variety of infections and other conditions that may appear singly, sequentially or simultaneously. Particularly common are infections caused by microbes that live inside cells and that depend on the cellular immune system to control them. Because the defences are deficient, many of the infections present in an atypical way and in a florid fashion such as to frighten the uninitiated. So far, treatment is at best palliative, and in the end, unavailing.

It must be remembered that most of the infections suffered by patients with AIDS are caused by microbes that are virtually or completely non-pathogenic for fit HCWs. Tuberculosis is an important exception. The appearance of tubercle bacilli that are resistant to several antimicrobials is a cause for concern. All HCWs who are to come into contact with AIDS patients should have been shown to be tuberculin positive, or have been vaccinated with BCG (Chapter 9).

FURTHER READING

See Appendix B.

Some tropical viral fevers

(Arboviruses, Arenaviruses, Filoviruses)

$\boxed{\textbf{22}}$

22.1 INTRODUCTION

Many of the common feverish illnesses of the tropics are caused by viruses that share common modes of transmission and have similar clinical presentations. There are too many individual diseases to be dealt with here one by one, so they are described in a general way, with examples selected from among the more important or notorious. The viruses that cause them may be divided into three groups. One is made up of agents loosely described as arboviruses, the others are the arenaviruses and the filoviruses. The speed of modern travel ensures that cases of diseases caused by these viruses can appear in temperate parts of the world. If they are to be diagnosed correctly it is important to bear the possibility in mind, and to ask patients exactly where they have been.

22.2 THE ARBOVIRUSES

The arboviruses were originally thought of as a group of related organisms transmitted between human or animal hosts by blood-sucking insects. The word is derived from the phrase *ar*thropod *bo*rne *virus*, where 'arthropod' more accurately describes what are otherwise loosely called 'insects'. Several hundred arboviruses have been discovered, of which perhaps 90 cause infections in humans. As the list has grown and as the modern system for the classification of viruses has developed the original definition of an arbovirus has become inadequate, but it is retained for convenience.

With important exceptions human arbovirus infections are zoonoses. The viruses survive in nature as the result of a continuous series of transmissions from one non-human vertebrate animal to an insect and then to another animal. Occasional human infections follow when an individual intrudes into this cycle and is bitten by an infected insect. In most cases infected humans cannot pass the infection on to another insect, so for these viruses humans are accidental or 'dead-end hosts', irrelevant to their continued survival in nature. The exceptions are dengue, urban yellow fever and sandfly fever. In these cases infected people are the sources of virus in what can develop into an epidemic based on human-insect-human cycles of infection.

In tropical areas the main insect vectors involved with the transmission of arboviruses are culicine mosquitoes. Ticks assume this role in cooler parts of the world where the mosquitoes fail to survive the winter. In some cases sandflies (*Phlebotomus*) and midges (*Culicoides*) act as vectors. Infections can also follow direct contact with blood or body fluids from an infected individual or animal. For some 'arbovirus' diseases this is the only route so far identified.

After an insect has taken a blood meal from an infected host any virus present enters some of its cells, and multiplies there. For this to happen the host's blood must contain at least one infectious dose of virus in each of the small volumes insects take for a meal. In many human arbovirus infections there is too little virus in the blood so that humans are often dead-end hosts for these viruses. The virus multiplies in the insect and after one or two weeks it spreads into its saliva. If they are not already immune the next animal or human that is bitten is then infected. Insects do not seem to suffer any ill effects from these infections, and the same is often true of the natural animal hosts of the arboviruses.

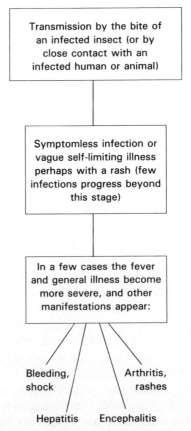

Figure 22.1 The development of illness in arbovirus infections.

Arbovirus infections in humans follow a discernible pattern, though with variations and some special features. Most of the infections are subclinical. Many more are mild and present in the form of indistinguishable short-term fevers. A small proportion progress to cause more characteristic and perhaps lethal illnesses. These patterns are illustrated in Figure 22.1. Many of the infections have geographical names (Kyasanur Forest, Ross River, West Nile) that indicate the places where the viruses were first isolated, or the ranges of their natural animal hosts and the insect vectors responsible for their transmission.

(a) Diagnosis, treatment and control

The diagnosis of arbovirus disease is usually serological, though virus isolation may be attempted. Treatment can only be supportive, though in severe infections modern intensive therapy saves lives. Control may be by vaccination when vaccines are available, or by an attack on the insect vectors. Vector control may be possible in villages and towns but this is not practical in the jungle or tracts of more open country. In these circumstances the vectors may be prevented from biting by suitable clothing, insect netting and repellents.

22.2.1 Dengue

This infection is endemic or occasionally epidemic in many parts of the tropics and subtropics. In endemic areas it is more common among children because the older members of the population have already been infected. Many infections are subclinical. A typical case of clinical disease presents with a high fever accompanied by acute pains in the muscles, especially those of the back and round the eyes, which are bloodshot (Box 22.1). A rash may appear on the third or fourth day. In more severe cases there is bleeding into the skin or from the nose and some patients then deteriorate rapidly with major haemorrhages and go into shock. Without intensive treatment dengue shock syndrome has a mortality that may reach 50%.

Box 22.1 'Dandy Fever'

The disease now called 'dengue' has been known for centuries under many different names. The present label seems to have appeared about 150 years ago in Mexico, where it may have been derived from a Spanish-American word meaning prudish or affected. If so it perceptively describes the mincing gait that painful muscles impose on sufferers. An old English name for it was 'dandy fever'. At one time the Germans called it *hundskrankheit* to highlight the bloodshot eyes that accompany the infection.

There are four distinct varieties of the dengue virus. Cases of dengue shock are more common in individuals who suffer second or third infections due to different varieties of the virus. The introduction of a new dengue virus into an area previously occupied by one of the others may be accompanied by an epidemic of serious disease. Dengue is spread from one human to another by mosquitoes. The most efficient and dangerous of these is *Aedes aegypti*, but fortunately this mosquito is relatively easy to control. Monkeys may also be infected. No vaccine is available. The fact that immunity to one dengue virus can be a disadvantage in an attack by a second suggests the need for caution.

22.2.2 Japanese encephalitis (JE)

JE is found in Asia and the Pacific area. In its most severe form it presents as an encephalitis with disorientation, stupor and coma. Some who recover have suffered permanent brain damage. The virus is maintained in nature by transmission between mosquitoes and birds or pigs. Humans, who are dead-end hosts for the virus, are infected by accident when they are bitten by the mosquitoes involved in these cycles. Large colonies of nesting water birds may provide the conditions in which enormous numbers of infected mosquitoes spill over to involve the local human population in an epidemic of the disease. A vaccine is available.

22.2.3 Yellow fever (YF)

YF was the first human disease in which a virus was identified as the cause, and the first virus infection in which an insect was implicated. It is confined to the tropical regions of Africa and the Americas (Box 22.2). Mysteriously it has never spread to the Indian subcontinent or to Asia. Initial fever,

---

Box 22.2 'Yellow Jack'

In the colonial era yellow fever (YF) caused a heavy mortality among Euorpean settlers, administrators and soldiers who were sent to the tropical areas of Africa and the New World. When the Spanish were driven out of Cuba, the occupying American troops began to suffer from the disease. In 1900 military volunteers were used in an experiment that showed YF could be spread by a mosquito, *Aedes aegypti*. When these mosquitoes were controlled, urban YF was eradicated on the island for the first time in 250 years.

In 1880 Ferdinand de Lesseps, fresh from his triumphant completion of the Suez canal, started to dig one on the Isthmus of Panama. He gave up in 1889 after 35 000 of his workers had died, many of YF. The work started again in 1904, and with urban YF now controlled the Panama canal was completed in 1914.

---

nausea and vomiting may progress to jaundice with bleeding and shock. In the most severe cases the liver is destroyed. Although the infection is often subclinical or mild there is a heavy mortality among those who develop the characteristic form of the disease.

In the wild the infection is maintained in a cycle that involves vertebrates, particularly monkeys, and the mosquitoes that live with them in the forest canopy. In Africa, where the disease originated, the monkeys have been exposed to the infection for so many generations that they no longer suffer from it. The disease was introduced into the Americas after the arrival of Columbus. New World monkeys have been exposed to YF for a comparatively short time and when infected they still fall ill. The disease has never spread to the tropical parts of Asia. The reason for this is unknown, but if Asian monkeys are deliberately infected with YF they develop a severe illness and most of them die (Box 22.3).

Box 22.3 Mortuary selection

When a new and lethal disease appears in a community it often seems that successive generations are progressively less susceptible to it. A good example of this is myxomatosis, a virus infection of rabbits.

The disease originated in S. America, where rabbits are commonly infected but suffer little from it. When the virus was introduced into Australia and Europe there was an explosive epidemic of an acute lethal infection that nearly wiped out the whole rabbit population. The few survivors were more resistant to the disease, and this resistance grew as generation followed generation. Rabbits breed very rapidly and in a few years it was already obvious that the disease was much less lethal than it had been.

The same process explains the differences in mortality from yellow fever between monkeys in different parts of the world. Humans breed more slowly so the action of mortuary selection is less easily followed, but it seems to have applied to syphilis in Europe and to tuberculosis.

People who enter the jungle may be bitten by mosquitoes that have acquired YF from monkeys. If not already immune they develop **jungle YF.** When they return home with the virus in their blood they may be bitten by *Aedes aegypti* mosquitoes that live and breed in and around their houses. These semi-domesticated mosquitoes are easily infected from human sources and can then transmit the infection to other people. If sufficient of the population of the village or town is not already immune an epidemic of **urban YF** develops. The 17D yellow fever vaccine gives a very high level of protection.

22.2.4 Sandfly fever

This is a mild but uncomfortable illness that resembles influenza but without the respiratory syptoms. The disease is spread between humans by sandflies. It is common in many of the drier parts of southern Europe, Africa and Asia. Small rodents may also be involved in transmission, and sandflies can pass the virus from one generation to another through their eggs.

22.3 ARENAVIRUS AND FILOVIRUS INFECTIONS

Viruses of these groups cause the rare but well publicized diseases Lassa fever, and Ebola and Marburg infections. All originate in Africa and have caused individual cases or outbreaks of disease, some of which have been severe. The infections resemble arbovirus diseases. Some are subclinical, others present with fever and muscle pains, and in more severe cases haemorrhages are common. The diagnosis is made serologically or by virus isolation. Transmission is by direct contact with infected individuals or animals, or with their body fluids. Ribavirin has been used in treatment and is helpful if given early. Because person-to-person transmission is possible, patients should be isolated, and specimens from them handled with care.

FURTHER READING

See Appendix B.

Rabies, pox and warts viruses, and prions 23

The viruses in the title of this chapter are brought together because they do not fit in anywhere else. The prions, a group of infectious agents of an unknown nature, are included for the same reason.

23.1 RABIES

The rabies virus is able to multiply in most warm-blooded animals. It causes symptoms when it invades nervous tissue and in most cases it is spread by bite. This is why its presence in saliva determines if a particular host can pass the virus on, or if it is a dead-end for it. The disease is maintained in nature by many wild animals such as wolves, jackals, skunks, squirrels, foxes and bats, for all of which (except the bats) it is highly lethal. Humans are infected when they are bitten by a four-legged animal that has the virus in its saliva. This period begins a few days before the animal shows any signs of illness, so an apparently normal animal can transmit the infection. Bats are different because they may have the virus in their saliva for a long time without any signs of illness.

Most human infections are the result of animal bites or licks on open wounds. A few cases have been reported in people who have visited bat-infested caves, or after corneal transplants from a donor who died of rabies, or as a result of laboratory accidents. Human infections become more common when the virus spreads into domestic animals, and especially when intimate pets like cats and dogs are involved. These may have been infected by contact with wild animals, or if the disease is poorly controlled, by spread from one pet to another. As human rabies is most often acquired from dogs the disease in this animal is described though, other than in bats, a similar pattern is seen in all of them.

After a dog has been bitten by another animal that is infected and has the virus in its saliva, there is an incubation period that lasts from 10 days to 6–8 months, most commonly 3–8 weeks. The first sign of illness is a change of temperament. An aggressive dog becomes friendly, a timid one, bold, a docile one, snappy and irritable. After some days the animal becomes excited and restless, and responds in an exaggerated way to noise and bright lights. At this stage the dog may run amok and bite everything and everyone it meets. This is furious rabies. Later the dog becomes stuperose and paralysed, with difficulty in swallowing. It creeps into a dark corner, sinks into a coma and dies. Sometimes an animal passes directly from an initial change of temperament to the last stage, without the furious phase. This form of the disease is called paralytic rabies. The rabies virus is present in the dog's saliva for the last two or three weeks of its life, but not for more than ten days before its illness has become obvious.

The incubation period in a human bitten by an animal with rabies is similar to, or shorter than in the dog. The first symptom is a change of sensation at the site of the bite, which may have healed. Next there is nausea, vomiting and a sore throat. The production of saliva increases and attempts to swallow cause painful spasms of the throat. An alternative name for rabies, hydrophobia, reflects the unwillingess of the victim to swallow water, or indeed anything. Recovery from established rabies is virtually unknown. The virus appers in human saliva so transmission from person to person is a theoretical possibility, but it has not been reported.

(a) Diagnosis, treatment and control

If possible the animal responsible for the bite should be caught and impounded alive in a safe place until expert advice as to the disposal has been received. Rabies is diagnosed in humans or animals by the microscopical examination of smears taken from the eye, or from the brain after death, or by culture of the virus. There is no treatment for the disease once it has appeared, but it is possible to vaccinate a person after they have been bitten, and before they become ill (Box 23.1). The long incubation period gives time for immunity to develop before the virus harms the central nervous system.

Box 23.1 Pasteur and rabies

At one time a person bitten by a rabid animal was taken to the local smithy where the wound was cauterized with a red-hot iron. As a child Louis Pasteur witnessed such treatment. Alternatively gunpowder was poured into the wound and set on fire. These extreme measures were preferable to the uncertain wait for the onset of a terrifying and ultimately lethal illness.

Later in life Pasteur infected rabbits with rabies and though he could not see the virus he knew it must be there. He had already found that virulent bacteria kept in unsatisfactory conditions lost their virulence, but could be used to immunize animals against one that was fully pathogenic. He dried the spinal cords of infected rabbits for various lengths of time. A cord dried for 14 days no longer caused rabies in a dog inoculated with it. Successive inoculations with cords dried for shorter periods immunized the dog so it could resist the injection of fresh fully virulent rabies virus. He repeated this experiment 50 times, without a single failure.

Modern vaccines are potent, and the 14 or 21 successive doses that were required with the older ones are no longer necessary. Immune globulin may be given with the vaccine, particularly if the attack by a rabid animal was very severe. People likely to come into contact with cases of rabies in humans or animals can be vaccinated before they are exposed to the disease. Travellers bitten by an animal should enquire, as a matter of urgency, if rabies exists in the locality. A few days lost before vaccination

Box 23.2 First rabies vaccinations

Rumour of Pasteur's successful experiments with dogs (Box 23.1) soon spread. One day in 1885 a human patient was brought to him. Joseph Meister aged nine had been savagely attacked by a mad dog. Though Pasteur knew that what he was about to do would be criticized, and what would happen if he failed, he took the risk and repeated his experiment on the child. Twelve inoculations were made, over ten days. The child survived. A little later Louise Pelletier aged ten was brought to him 37 days after she had been bitten by a rabid dog. Against his better judgement Pasteur yielded to the pleading of the girl's parents and she was vaccinated. It was too late, and she died of rabies. Public opinion ebbed and flowed, and for a time there was a chance that Pasteur would be accused of manslaughter, but eventually his vaccine was accepted.

In later life Joseph Meister became doorkeeper at the Institut Pasteur in Paris. He committed suicide in 1940 because he could not bear the thought of the German occupation of France.

is commenced may be critical (Box 23.2). Vaccines are also available for animals.

Control is exercised in a variety of ways. In countries where there is no rabies in the wild animal population, strict control is imposed on the importation of any that might be infected. This involves their quarantine for a period, usually six months, so they can be observed for the onset of any illness that might be rabies. In countries where rabies is already present among wild animals, control depends on the management of the wild population, the registration and vaccination of domestic animals, and the control of strays.

23.2 THE POXVIRUSES

23.2.1 Smallpox

The last natural case of smallpox was diagnosed in a hospital cook in Somalia in 1977, though there were other cases as a result of a laboratory accident in 1978. The World Health Organisation declared the disease extinct in 1979. So far as is known the only smallpox viruses that still exist are kept under lock and key in two laboratories in the world (Box 23.3)

Box 23.3 Smallpox entombed?

The smallpox virus is one of the few microbes that can withstand desiccation. This suggests that in an unusually dry and cold grave the smallpox virus might still survive. The mummy of the Pharaoh Rameses V has a rash, probably of smallpox. A person who dies of smallpox is likely to have an extensive rash that contains billions of virus particles.

Between 1729 and 1858 the crypt of Christchurch in Spitalfields, London was used for interments. At the time smallpox was common in London. Over 100 years later it was decided to use the crypt for other purposes. The exhumations provided an opportunity for an archaeological survey. In April 1985 one of the corpses examined was found to have a rash. A structure that resembled a smallpox virus was seen by electron microscopy in material taken from the rash but no virus was grown. It is possible that somewhere a viable smallpox virus is entombed, perhaps in a body buried in arctic permafrost.

23.2.2 Other poxviruses

The human smallpox virus is a member of the large group of poxviruses, each of them associated with one type of animal. The monkeypox virus may cause a disease in humans that resembles mild smallpox, but the infection does not spread easily from person to person. The poxviruses of other animals may also cause human infections, but when they do the disease is usually mild and limited to the site of inoculation. The cowpox virus used by Jenner to protect against smallpox is a good example. The vaccinia virus used more recently as a smallpox vaccine is not the same as the cowpox virus, and it is not known where it came from. In the early days of vaccination 'lymph' was collected from the local reaction that developed on a vaccinated child. This was then used to vaccinate the next batch of children (Box 23.4). In the course of thousands of such transfers it is likely that the virus underwent several mutations, or was mixed with another from somewhere else.

Box 24.4 Smallpox vaccination as it was

Smallpox vaccination was made compulsory in the UK in 1853. The same thing happened in the USA in 1855 though the rule was not enforced until 1872. After vaccination in the UK a certificate was provided that declared:

> Every . . . person having custody of a child shall take it . . . on the same day in the following week to the Public Vaccinator . . . (to) ascertain the result and if he sees fit, take from the child lymph for the performance of other vaccinations . . .

Non-compliance attracted a fine 'not exceeding Twenty Shillings'. Vaccination with 'humanized lymph' ceased in 1899.

Vaccination reduced the annual mortality from smallpox in the UK from about 3000 per million of the population a year before it was introduced to about 300 in the mid- 19th century and to 13 per million by 1900.

Orf is a disease of sheep and goats caused by another of the poxviruses. It may be transmitted to humans (often shepherds) to cause a large weeping nodule, usually on the hands.

Molluscum contagiosum is a skin disease due to a human poxvirus that is quite distinct from the smallpox virus. Infection causes wart-like nodules on the skin that have smooth surfaces and perhaps a central depression. The lesions often appear in succession, each nodule lasting for two or three months. Transmission may be sexual, or be the result of some other close physical contact.

23.3 THE WARTS VIRUSES

There are more than 60 different types of warts viruses (papillomaviruses) that may cause infections in humans. The different types tend to be associated with certain parts of the body. Common warts, often on the hands or knees, and plantar warts (verrucas) are well known. They may last a long time, be multiple and troublesome. Genital warts are usually sexually transmitted and are more fleshy. They are found in the moist areas around the genitalia, and on the uterine cervix. Warts sometimes develop on the vocal cords. These laryngeal papillomas and the genital warts may undergo malignant change. It is not known if the virus is the cause of the malignancy, or merely predisposes to it.

Common warts usually disappear on their own after a variable, perhaps long, period. Whatever treatment happens to be in vogue at the time, from a rub with granny's gold ring to the kiss of a frog, gets credit for the cure. Warts may be destroyed by freezing with dry ice or liquid nitrogen, or they may be 'vaporized' in the beam of a laser. The latter does not inactivate the virus, however, which may be inhaled by those in the vicinity of the operation as it is performed.

23.4 THE PRIONS

'Prion' is a name proposed for members of a group of infectious agents of an unknown nature. Their existence has been inferred from the transmission of certain diseases from animal to animal. If an extract of brain tissue from one with the infection is injected into another the disease is reproduced, and so on, in series. Whatever is responsible must multiply in each host for such serial transmission to be possible, so the disease cannot be due to a simple poison.

Prions have not yet been identified with certainty by electron microscopy (Ozel & Diringer, 1994). They are still infectious after treatment by heat or strong disinfectants that destroy all other microbes, so they must have a most unconventional structure. A they can only be grown in experimental animals little progress has been made in the study of the diseases they cause. Prions evidently multiply very slowly because the infections have very long incubation periods. Infection does not produce any immunity.

Prions cause disease in humans and animals. Symptoms arise when they multiply in the cells of the central nervous system. A characteristic protein appears and multiple small holes develop in the substance of the brain. Because of this the prion diseases so far identified have been classified as spongiform encephalopathies.

The prion disease scrapie has been recognized among sheep and goats in Europe for hundreds of years. Infected animals suffer from an itch that makes them rub themselves against trees and posts. More recently prion diseases have appeared in other animals, including mink bred in captivity for their fur, domestic cats and in cattle (bovine spongiform encephalopathy, or mad cow disease). This extension to other animals may have been caused by the inclusion of parts of slaughtered sheep, including the brain, in commercially-produced animal food.

Diseases in humans caused by prions have been described under three names, though they may all be the same. The most widely distributed is the rare Creutzfeldt-Jakob disease (CJD) that is found with varying frequency in all parts of the world. The disease presents as a progressive dementia that, once it has appeared, leads to death within a year. Variations in the frequecy of the disease may be associated with different levels of genetic predisposition among people with common heritages. When the disease runs in families it has been called the 'Gerstmann-Straussler syndrome'. Kuru was a prion disease that afflicted members of the Fore tribe in Papua New Guinea. It is thought that it killed upwards of 10% of the population. Transmission was the result of ritual cannibalism. This is no longer practiced, and kuru has disappeared. It has been suggested that the disease was originally introduced when a missionary who was infected with CJD was eaten.

The cause of CJD is unknown, though the examples of animal prion diseases and kuru suggest that the route of infection might be by mouth. Against this must be set the fact that people have eaten sheep and goat meat for millennia. No association has been noted between this practice and CJD, though difficulty with diagnosis, the long incubation period and differences in the acceptance of brain as an item of diet lessen the impact of the assertion.

What is much more certain is that some medical treatments have been the cause of significant numbers of cases of CJD, and that there will be more of them. So far the largest number have been in people given growth hormone for the correction of short stature, or gonadotrophin for infertility. At one time these hormones were extracted from pituitary glands that had been removed at post-mortem examinations. Some of these clearly came from people who had died of CJD, or who were incubating it when they died of something else. The hormones are now made more safely by methods that involve genetic engineering. Other cases of CJD have followed corneal transplantation, and neurosurgical operations. These infections have been the result of the transfer from one person to another of material from the brain, or from the eye that is anatomically a part of it. The incubation periods in these iatrogenic cases of CJD have been between one and 20 years.

Neurosurgical equipment treated by conventional methods of sterilization is not sterile with regard to prions. Special methods such as autoclaving for three times the normal period of exposure are required when dealing with apparatus that has been used on patients who suffer from CJD. The long incubation period means that some people are subjected to neurosurgery or are used as donors of tissue for grafts before their illness can be recognized.

FURTHER READING

1. Ozel, M. and Diringer, H. (1994) Small virus-like structure in fractions from scrapie hamster brain. *Lancet*, **343**, 894–5.
2. See Appendix B.

Fungi, and the Infections They Cause

Fungal infections 24

24.1 INTRODUCTION

Fungi differ from bacteria in a number of ways. For example, they possess true nuclei and they may be classified as protists to distinguish them from animals and plants. They often grow in the form of long, branching filaments called **hyphae** to produce a tangled felted mass, the **mycelium** (Figure 24.1). From this the words used to describe the study of the fungi and the diseases they cause, **mycology** and the **mycoses**, are derived. Mushrooms in the fields, bracket fungi on tree-stumps, the moulds that develop on stale bread or cheese, or on a damp wall, are mycelial fungal colonies. Other fungi grow as microscopic Gram-positive oval structures called **yeasts** (Figure 24.2). These are larger than bacteria, and they multiply by **budding**. A bud on the side of a parent cell grows until it is large enough to break away as a new daughter cell. Under certain conditions yeasts develop into mycelial forms and so justify their classification among the fungi.

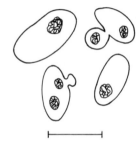

Figure 24.2 A yeast-like fungus, budding to form a daughter cell. (The bar is 10 μm in length.)

The cell walls of fungi contain materials like chitin rather than the peptidoglycan of bacteria. For this and other reasons antibacterial drugs are of

Fungal spores

Mycelium on a
solid surface

Hyphae

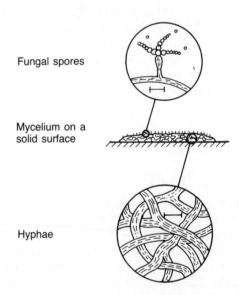

Figure 24.1 A fungal mycelium with matted hyphae and one form of the apparatus that distributes fungal spores into the air. (The bars are 10 μm in length.)

no use against them, indeed many of the drugs used in the treatment of bacterial infections originally came from fungi. Penicillin was at first extracted from the mould *Penicillium notatum* (Box 24.1). An enormous number and variety of fungi exist in nature, ranging from some that are edible to microscopic pathogens that prey on other living things. The filaments of some of the fungi that grow in the soil spread out for distances that can be measured in kilometres. In the terms of area covered these are the largest living things on earth.

Box 24.1 Penicillin

European folklore suggested that a paste of mouldy bread should be applied to a wound to prevent infection. In his bacteriological experiments Joseph Lister noticed that bacteria sometimes failed to grow in the presence of a fungus. The credit for the discovery of penicillin, however, has gone to Alexander Fleming. In 1928 in London he isolated a fungus that was responsible for a major antibacterial effect. The enormous importance of this discovery was not immediately apparent to Fleming or to his colleagues, and it was another 10 years before Howard Florey and Ernst Chain isolated the active substance. As a result of this new work penicillin emerged as the first antibiotic drug (Chapter 30).

Mycologists classify fungi according to the type of sexual reproduction they undergo. Until recently no sexual stage had been observed for most of the human pathogens and they were classified as *fungi imperfecti*. This is no longer true and many pathogenic fungi have been reclassified and have been given new names. The older terminology is still used in most medical texts.

Some of the fungi pathogenic for man and animals can only grow in the horny keratinized tissues of the surface layer of the skin and in the hair and nails. The diseases they produce are called the **superficial mycoses**. Other pathogenic fungi are able to invade and spread through the tissues to cause the **deep mycoses**. A third group of infections are caused by fungi widely distributed in nature that are usually harmless to humans, but which may cause serious infections in debilitated or immunodeficient patients. The diseases they cause are the **opportunistic mycoses**.

24.2 THE SUPERFICIAL MYCOSES

(a) *Pathogenesis*

The superficial mycoses appear as **ringworm** or **tinea** and the fungi responsible are called **dermatophytes**. The infections may be acquired by contact with infected animals or people, or directly from the soil. Dermatophytes depend on the digestion of keratin for their nutrition so they can only grow where this substance is found, in the most superficial part of the skin and in the hair and nails. Sometimes when they grow in these dead tissues they may cause no symptoms, or they can produce disease when their waste or other products diffuse into neighbouring, living parts of the body. When this happens in an area of skin that is dry the inflammatory response causes itching and a rash with the formation of vesicles. The horny layer of the skin thickens and when viewed from above the whole lesion has a circular or wavy outline. As the infection spreads outwards there is a tendency for the part in the centre to heal and the appearance produced has suggested the common name 'ringworm'. In the moist areas of the body the thickened skin is soft and tends to split to produce the cracks seen in athlete's foot. In these moist areas secondary bacterial infection is an important part of the pathology.

When hair is infected it is weakened and breaks off close to the skin to produce the bald or 'mousey' appearance that may be called mange in animals. Infected nails are deformed, discoloured and brittle.

When dermatophytes grow in the skin their hyphal filaments spread out like the roots of a plant. The older hyphae break up into infectious arthrospores that are shed with skin squames, perhaps to infect others. Patients may develop an allergy to fungal products that causes an eruption on the hands. This allergic reaction does not contain the fungus, so treatment with antifungals is of no use. It is necessary to find the primary fungal infection (perhaps not very obvious) and treat that. Infections may be of the feet (tinea pedis) with extension to the groin (tinea cruris), or of the body (tinea corporis, ringworm) or of the head (tinea capitis, ringworm of the scalp).

(b) Diagnosis and treatment

Specimens of affected hair or skin taken from the advancing edge of a lesion, or of damaged nails, are examined under the microscope. If hyphae or arthrospores are found this indicates the presence of a dermatophyte infection. Fungi can be grown in the laboratory on special media and identified precisely, but this may take three weeks or more. Treatment consists of the removal of dead tissue, attention to any secondary infection, and the prolonged application of an antifungal agent. In severe cases they may be given by mouth.

24.3 THE DEEP MYCOSES

The deep mycoses are caused by fungi whose normal habitat is the soil or decomposing organic matter. It is convenient to divide the diseases they cause into subcutaneous and systemic forms, according to how widely and deeply they spread in the tissues.

24.3.1 The subcutaneous mycoses

One of these infections (sporotrichosis) follows the implantation of the causative fungus (*Sporothrix schenckii*) into the subcutaneous tissues. The source of the fungus is usually the soil and the point of entry often on the hand. If not eliminated the fungus multiplies and spreads slowly along the lymphatic channels. These thicken, and nodules or abscesses develop at intervals, like knots on a piece of string. The infection is treated with potassium iodide.

Sometimes the disease is due to other fungi that spread from the point of implantation to penetrate into and through local tissues, eventually to reach bone. A chronic, swollen fibrotic and destructive lesion called a **mycetoma** develops, with multiple sinuses from which pus is discharged that contains coloured grains of fungal mycelium, visible to the naked eye. The condition usually involves the feet of people who go barefoot in tropical areas. Some actinomycetes cause a similar condition (Chapter 7 and Box 7.3). A number of different fungi may be involved. Diagnosis is by microscopy and culture of tissue or of the grains discharged from the mycetoma. Treatment of true fungal as distinct from actinomycotic mycetomas is difficult, and radical surgery is usually necessary.

24.3.2 The systemic mycoses

The fungi that cause the systemic mycoses are found in abundance in the environment in particular geographical locations. Sooner or later most members of the human population who live in these places are infected. Most of the infections are subclinical or very mild, though a few individuals develop severe or fatal illnesses. These more severe generalized infections usually progress slowly, spread through the lymphatics, and produce a fibrotic tissue reaction. Infection induces hypersensitivity, and as the outcome depends on the efficiency of the patient's cellular immune system the diseases resemble tuberculosis (Chapter 9). The fungi grow in the soil to form mycelia and produce infectious spores. In human tissues they multiply in the form of spore-like yeasts.

One of these diseases is histoplasmosis, caused by *Histoplasma capsulatum*. This fungus is found in the soil in several parts of the world, particularly in parts of the eastern USA (Figure 24.3), and in Asia and Australia. It grows particularly well in soil manured with chicken or bat faeces. Infection follows

Figure 24.3 In the USA the infection, histoplasmosis, is common in the Mississippi basin. In the area marked on the map 50% or more of the population have positive histoplasmin tests, indicating that they have been infected by *H. capsulatum* at some time in their lives. Nearly all of these infections are symptomless.

the inhalation of dust containing spores of the organism. Illness, when it occurs, varies from an acute benign self-limiting respiratory infection, through a chronic form, to an acute disseminated and rapidly fatal disease rather like miliary tuberculosis.

Coccidioidomycosis is caused by *Coccidioides immitis*. This fungus is found in the soil in the southwestern part of the USA and in Latin America. The infection it causes is similar to histoplasmosis. *Blastomyces dermatidis* is the cause of blastomycosis, one of the less common systemic mycoses.

(a) Diagnosis and treatment

The diagnosis of any of the systemic mycoses is made by microscopy of tissue, sputum or other material. A search is made for the characteristic spore-like cells of the fungus. The diagnosis may also be made by culture and serological examination. As with tuberculosis intradermal tests are available to detect the hypersensitivity that results from infection. For histoplasmosis the reagent is called histoplasmin (Figure 24.3).

24.4 THE OPPORTUNISTIC MYCOSES

In healthy individuals the fungi that cause these diseases are weak or non-pathogenic but they may cause serious disease in those whose immunity is defective. They may be acquired by self-infection from an individual's own normal flora or by cross-infection from a neighbour, or from the environment.

24.4.1 Candidiasis

This infection is caused by *Candida albicans* and a few related species of yeasts. These fungi are often found in small numbers as a part of the normal flora of the body. If something happens that allows them to multiply in an uncontrolled way on the surface of the body, they cause soreness and perhaps itching in the mouth, anus or vagina, with a discharge from the latter sites. This may be a complication of the use of broad-spectrum antibacterial drugs that upset the balance of the normal flora. Patients may be predisposed to such overgrowth by diabetes, debility, and immunodeficiency (notably in AIDS), indwelling urinary or intravenous catheterization, intravenous drug abuse and steroid treatment.

Candidiasis typically involves mucous membranes and produces the white adherent patches commonly seen in 'thrush' in the mouths of infants. Candida may also cause infections of nails and nail beds, and may infect skin in the moist areas of the body. More severely immunodeficient patients can suffer from serious generalized infections. These involve the lungs, kidneys, or the heart, as a part of a candidal septicaemia.

The diagnosis of candidiasis is made when the yeasts are seen by microscopy in specimens taken from the affected part, or are isolated by culture. Care must be taken when positive reports are interpreted as small numbers of yeasts are often found as a part of the normal flora of the body. The antifungal drug nystatin that can be used to treat local infections is not absorbed when taken by mouth. Some newer antifungal drugs that are absorbed when given by this route may also be injected. They are more often used in severe or generalized infections. If life is threatened the rather toxic antifungal drug amphotericin B can be employed.

24.4.2 Cryptococcosis

Cryptococcus neoformans is a yeast-like fungus with a notably thick capsule. It is widely distributed in the environment, and may be found in the droppings of pigeons. Human disease due to *C. neoformans* is rare, though it is found increasingly in patients with AIDS. The disease seems to start with a subclinical pulmonary infection that progresses in the immunodeficient to produce fungaemia and meningitis. The meningitis usually runs a chronic course and is fatal if not treated.

The diagnosis is made when yeasts with characteristic thick capsules are found in the cerebro-spinal fluid (CSF) or in other clinical material. The capsular substance may be identified serologically in CSF, even when the yeast cannot be seen because of prior treatment or because the infection is localised. Cryptococci can be grown in the laboratory.

24.4.3 *Pneumocystis carinii*

Although this parasite was classified as a protozoon, recent examination of its nucleic acid has suggested that it is a fungus. The disease it causes resembles an opportuistic mycosis but if *C. carinii* is a fungus its mycelial form has yet to be recognized.

In individuals whose cellular immune system is deficient, this parasite invades the lungs to cause severe or fatal pneumonia. At one time this was a rare disease of weakly infants or unusually debilitated children or adults. It is now more common because of the use of immunosuppressive drugs in transplantation, and because of the AIDS epidemic (Chapter 21). The diagnosis is made by the microscopical examination of specially stained material collected by bronchoscopy. Expectorated sputum is rarely of use. Treatment is with cotrimoxazole or pentamidine, and cotrimoxazole may be used prophylactically in patients predisposed to the infection.

24.4.4 Other opportunistic mycoses

Large numbers of spores of many fungi are regularly present in the air so everyone inhales them every day. In a few people they cause allergies such as asthma or hay fever but for most individuals they are harmless. In debilitated patients, particularly if they also have an anatomical abnormality, these otherwise non-pathogenic fungi may gain a foothold and establish an infection. An old tuberculous lung cavity in a poorly controlled diabetic can become the focus of an infection with, for example, a member of the genus *Aspergillus*.

A patient who is immunosuppressed as a part of the treatment of leukaemia or after a transplant may become the victim of a severe progressive infection with this or some other mould. The diagnosis is made by microscopical and cultural identification of the fungus. Treatment is often difficult and may fail.

FURTHER READING

See Appendix B.

Infections and Infestations Caused by Other Parasites

The protozoa | 25

(*Plasmodium, Trypanosoma, Leishmania, Toxoplasma,* Amoeba, *Trichomonas, Giardia, Balantidium, Cryptosporidium*)

25.1 INTRODUCTION

Although pathogenic bacteria, viruses and fungi are also parasites the term parasitology has by custom been reserved for the study of the protozoa, the helminths and those insects and insect-like creatures that may cause pathology. Protozoa are true microbes and, although helminths and insects are not, it is convenient to describe them as part of the study of microbiology (Chapters 26 and 27). The common human pathogenic protozoa are listed at the head of this chapter.

Human pathogenic protozoa exhibit a wide variety of structures and lifestyles. They may live in the tissues or on the surfaces of the body in the gastrointestinal and urogenital tracts, and some inhabit both deep and superficial sites at different times or stages of their development. Some are transmitted by the faecal-oral route, some sexually, and in the warmer parts of the world many of the more important are transmitted by insects. Some are purely human parasites, and others are shared with a variety of animals. They include the causes of several major infectious diseases. A great deal is known about this important group of pathogens. What follows is no more than a synopsis.

25.2 MALARIA

(a) Pathogenesis

It is not bad air from the marshes that causes 'mal-aria', but the mosquitoes that breed in them (Box 25.1). To be more precise the infection is spread by female anopheline mosquitoes. To do this each mosquito must at some time have bitten someone who suffers from malaria. With this meal of blood she acquires male and female forms of the malaria parasite which unite

Box 25.1 The ague

Alexander the Great died of malaria in 323 BC, in Babylon. For centuries the 'miasma' from the Campagna marshes near Rome killed the local peasants and frightened the citizens of the Imperial City. The Sinhalese Kingdom of Sri Lanka fell as a result of malaria. The disease was recognized by the disciples of Hippocrates and many of the agues of Victorian England were caused by it.

In 1880 a French army surgeon Alphonse Laveran saw the malaria parasite in the blood for the first time. Ronald Ross, a British army surgeon, proved that malaria is transmitted by mosquitoes, but it was not until 1948 that the life cycle of the parasite was completely worked out when the stage in the liver was discovered.

inside her to produce multiple offspring. When these reach the mosquito's mouthparts they are injected with her saliva into the donor of her next blood meal. In the new human host they multiply again, at first in the liver, and then in the blood. Here more sexual forms are produced, ready for the next mosquito that comes along to complete the life cycle of the parasite. The protozoa that cause malaria belong to the genus *Plasmodium*.

Not satisfied with the completion of its life cycle in this way the parasite also multiplies asexually in the red blood cells of its human host. When it does so it destroys them, and this dictates the nature of malaria. As each crop of parasites reach maturity the cells that contain them break open. The parasites released invade new red cells, and the process is repeated (Figure 25.1). After a time these destructive cycles tend to become synchronized so that large numbers of cells break down more or less together to cause a sharp

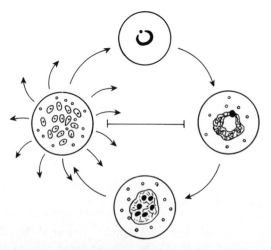

Figure 25.1 The asexual multiplication of the malaria parasite *Plasmodium vivax* in human red blood cells. With this species granules called Schuffner's dots also appear inside the cells and the cycle is repeated every 48 hours. (The bar is $10\,\mu m$ in length.)

attack of fever. Each cycle takes 48 or 72 hours to complete, to give rise to tertian (*P. vivax*, *P. falciparum*, *P. ovale*) or quartan (*P. malariae*) forms of the disease, with a bout of fever repeated every third or fourth day (the first day of fever is counted as 'one'). This characteristic periodicity is only seen in well established cases of malaria, so is of no diagnostic value in an early infection.

Episodes of fever start with a chill, accompanied by shivering. The individual then becomes hot with headaches, muscle pains and perhaps vomiting. Each attack ends with a drenching sweat. The destruction of red blood cells leads to a progressive anaemia, debility and an enlarged spleen as this organ works overtime to clear away the debris. In the presence of malnutrition the anaemia is more severe, and may be fatal. A recent estimate suggests that 1–1.5 million people die as a result of malaria each year in Africa alone, and that it is the cause of perhaps twice as many deaths in tropical regions worldwide (see Appendix 2, entry 13).

Because of its severity the disease caused by *P. falciparum* is called 'malignant tertian malaria'. Red blood cells that contain this form of the parasite congregate in the internal organs, where they block the capillaries. In the brain this causes cerebral malaria. Patients with this condition complain of severe headaches and neck stiffness before they sink into a coma. In the absence of early, vigorous treatment death soon follows. Blackwater fever is another complication, perhaps precipitated by treatment. A massive destruction of red blood cells releases haemoglobin into the blood. This spills out through the kidneys into the urine, which appears dark as a result.

(b)   Diagnosis, treatment and control

A laboratory diagnosis of malaria is made by a microscopical examination of the blood. This also allows the species of plasmodium responsible to be identified. For many years the drug chloroquine was the mainstay of treatment but strains of the parasite resistant to it and to several other antimalarials have appeared, and are now common in some parts of the world. These resistant infections may have to be treated with quinine, the original antimalarial drug extracted from cinchona bark (Box 30.2). A new less toxic alternative is artemisinin which is also extracted from a plant, in this case, wormwood (Box 25.2).

Box 25.2 Wormwood

A new antimalarial drug has been extracted from a member of the genus *Artemesia*. This genus includes the herb tarragon among about 400 species of shrubby plant. An extract of *A. absinthium* ('wormwood') was used to cause diarrhoea and expel worms from the gut, and as its Latin name suggests it also provided a flavour for absinthe, still used in vermouth. For thousands of years extracts of several species have been employed for medicinal purposes. An annual variety (sweet wormwood or *A. annua*) was used in China to treat conditions as different as haemorrhoids and fevers. It must have worked sometimes because recent research has shown that an extract known as *qinghaosu* (or artemisinin) is a powerful antimalarial agent. Some synthetic derivatives of the natural product are even more active (Hein and White, 1993).

Malaria can be controlled by attention to its vector, the anopheline mosquito, and by the use of an antimalarial drug that is taken regularly, as a prophylactic. Attempts have been made to develop a vaccine for malaria, but so far these have been unsuccessful. Among prophylactic drugs proguanil was ideal as it was effective, of low toxicity, and of no use in the treatment of malaria. Unfortunately the valuable therapeutic drug, chloroquine, was also widely used as a prophylactic. The predictable result was the emergence of chloroquine resistance, and this is now a serious problem.

Mosquitoes may be controlled in several ways. They can be prevented from biting by the use of mosquito netting, suitable clothing and repellents. Alternatively their breeding may be interfered with, or they may be killed. Mosquitoes require water in which to breed. If this is covered, or contains fish that eat the larvae, some control is possible. At one time the insecticide DDT (dichloro-diphenyl-trichlorethane) killed mosquitoes so successfully that there were hopes that malaria might be eradicated, and this was achieved in some places, for a time. Regrettably mosquitoes became resistant to DDT, which was also unexpectedly toxic. Other insecticides are available but none combine the essential qualities of DDT with greater safety.

Malaria is still out of control in many parts of the world, and it remains a most serious problem. Travellers to these areas need to be aware of this, and to take advice on the prophylaxis they need. Every year a number of them die of malaria because they have failed to take their prophylactic drugs, or have used the wrong one. When the traveller returns home and falls ill the correct diagnosis may not be considered until, in the case of malignant tertian malaria, it is too late.

25.3 AFRICAN TRYPANOSOMIASIS OR SLEEPING SICKNESS

Human sleeping sickness is found in a belt that extends across Equatorial Africa. This corresponds to the distribution of its vector, the tsetse fly. Trypanosomes are responsible (Figure 25.2) of which there are two varieties. One, *Trypanosoma brucei rhodesiense*, is the cause of a rapidly progressive disease that is fatal in weeks to months. It is found chiefly in East Africa. The other, *T.b. gambiense*, is the cause of a less severe disease that is found more often in West and Central Africa. This takes several years to run its course. Trypanosomes multiply at the site of the bite of an infected tsetse

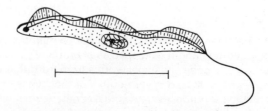

Figure 25.2 A trypanosome. The central nucleus is visible and, at one end, a parabasal body. Arising from this is a flagellum that is attached to the tryapnosome for much of its length, to form an undulating membrane. (The bar measures 10 μm.)

fly and then spread widely to cause a generalized illness with fever headache and enlarged lymph nodes. Eventually the central nervous system is invaded. The patient becomes wasted and sinks into coma ('sleeping sickness'). Without treatment the disease is fatal. *T. b. gambiense* is largely restricted to the human race, while *T. b. rhodesiense* also infects animals, including domestic cattle.

Trypanosomes evade the body's immune system by changing their coats, with a wardrobe of about 1000 variations available to them (Chapter 3). The diagnosis is made when trypanosomes are seen in the blood, in lymph nodes, or in cerebro-spinal fluid by microscopy, or serologically. A number of trypanocidal drugs have been used in treatment, or for prophylaxis. If treatment is delayed until after the brain has been damaged, recovery is not complete. Control measures are largely directed at the vector, the tsetse fly. This has been quite successful, though in some localities breakdowns due to political disturbances have led to the re-emergence of the disease.

25.4 AMERICAN TRYPANOSOMIASIS OR CHAGAS' DISEASE

Disease due to *T. cruzi* is found in Central and South America. It is spread by reduviid cone-nosed or kissing bugs that infest badly constructed and poorly maintained homes. They cannot fly and grow to 2.5 cm (an inch) in length. *T. cruzi* is a widespread parasite of many animals including dogs, cats, rats and mice. It multiplies in the bug's intestine, and passes out in its faeces. When a child is bitten its skin is contaminated with the faeces, and the parasite enters the body through the bite or some other abrasion. After this there is a brief illness, but the principal disease appears later in life, with a progressive attack on the heart or the gut. The most common cause of death is heart failure. In this chronic form of the infection the parasites live inside cells and the disease is difficult to diagnose and resistant to treatment. Control is directed at the elimination of the vector bug.

25.5 LEISHMANIASIS

There are may species of *Leishmania* that may cause infections in humans. They differ in their geographical distributions and in the diseases they cause. They are spread by the bites of infected sandflies. The parasites may be found in animals as well as in humans and the sandflies may acquire their infections from either. The diseases caused by leishmanias are of two types. First are a group of superficial infections that involve skin and mucous membranes, and second, a systemic form in which the viscera are attacked. The parasite multiplies within the cells of its host, so to some extent evades the body's immune system. Both types of disease are found in tropical and subtropical areas of the world.

25.5.1 Cutaneous leishmaniasis

After an incubation period that may last for weeks or months a papule develops in the skin at the site of the bite of an infected sandfly. The papule develops into an indolent ulcer that, left to itself, may take months or years to heal. In the Americas the infection may flare up several years later as a progressive destruction of the nasopharynx. This is called espundia or mucocutaneous leishmaniasis. In the Mediterranean area, Middle East, Asia and Sub-Saharan Africa the disease is less aggressive and it goes under a variety of colloquial, geographical names such as Oriental Sore or Baghdad or Delhi boil.

25.5.2 Visceral leishmaniasis

This is a much more serious disease, with fever, enlargement of the liver, spleen and lymph glands, and a severe progressive anaemia. The incubation period, measured in months, is followed by an increasing loss of weight. The disease, which is usually fatal if not treated, is found in restricted areas within the general distribution of cutaneous leishmaniasis.

(a) Diagnosis, treatment and control

Both forms of the disease are usually diagnosed by the microscopical examination of affected tissues. Cultural methods and serological tests are also available. When required treatment is with rather toxic antimonal drugs that must be given for long periods. Control measures are directed at the sandfly and their animal hosts, so the methods vary with the locality concerned. It has been noted that the visceral disease becomes less common in places where the population of dogs has been reduced. Sandflies are very small. If they are to be excluded from buildings or beds the size of the holes in the netting that must be used makes it oppressive in hot climates, though the design of modern air-conditioned buildings overcomes this problem.

25.6 TOXOPLASMOSIS

Toxoplasma gondii is an extremely common protozoon parasite of humans and other vertebrates. Cats and related animals are the key hosts as the sexual stage of the parasite's life cycle can only be completed in the feline intestine. Eggs (oocysts) pass out in their faeces and remain viable in the soil for up to a year. If they are eaten by other animals these oocysts hatch and the parasites that emerge escape from the gut and invade the tissues to cause toxoplasmosis. This process more often involves small numbers of parasites so is nearly always symptomless, but the host develops antibodies and the parasites develop into cysts in various parts of the body. They remain viable for life. The cycle is complete when cats eat animals with these cysts in their flesh. Non-feline animals including humans are also infected when they eat flesh that contains the cysts, though they cannot support the intestinal phase of the life cycle. Humans acquire the disease by eating raw or undercooked meat, usually mutton or pork, less often beef. Milk from infected goats or cattle may also transmit it and children may be infected if they play in soil or sand contaminated with cat faeces.

Toxoplasmosis presents as a clinical disease if unusually large numbers of parasites are consumed at one time, or if dormant cysts, activated by immunodepression, begin to multiply. The infection resembles glandular fever (Chapter 20). Toxoplasmosis has become prominent for two reasons. First, if a woman is infected with *T. gondii* early in pregnancy her fetus is likely to be involved. It may die, or develop permanent damage to the brain, eyes or liver. Infections later in pregnancy cause less severe injuries. The second reason is that an individual who is immunosuppressed can suffer from a reactivation of any dormant cysts of *T. gondii* that may be present, with severe consequences if the parasite begins to multiply in the brain. This can happen in AIDS. A primary infection in an immunosuppressed person sometimes develops into an overwhelming generalized disease. This can follow if a transplanted part carries the parasite to a recipient who meets it for the first time.

The laboratory diagnosis is made serologically, and culture may be attempted. A number of drugs are available for the treatment of toxoplasmosis, when this is necessary. Some of them may be used prophylactically to prevent the reactivation of disease in patients with AIDS. For prevention all meat should be cooked until it has changed colour. Cats should also be fed food that has been cooked, and they should be discouraged from hunting. Children should not play in earth or sandpits frequented by cats. Unless it is known that they have already been infected, pregnant women should be particularly careful to wash their hands after they have handled raw meat, and they should limit their contacts with cats.

25.7 AMOEBIASIS

25.7.1 *Entamoeba histolytica*

This amoeba is an entirely human parasite. It lives in the large intestine in an active, motile form, called a **trophozoite** (Figure 25.3). Trophozoites crawl over the surface of the mucosa of the bowel, where they feed and multiply. Occasionally they develop into cysts that pass out in the faeces and survive for a long time in the environment. When these cysts contaminate food or drink they re-enter the human gut where trophozoites re-emerge to complete the life cycle. Carriage is usually asymptomatic and in some tropical and subtropical regions, particularly where human faeces is used for manure, half the population may be involved. In other parts of the world carriage is more restricted and the people concerned tend to live in situations where faecal-oral transmission is more likely, as in families, mental institutions and among male homosexuals.

In certain circumstances amoebae may cause pathology. The reason for this is not clear, but some amoebae are recognizably more pathogenic than others, and host factors must also play a part. Disease begins when the parasites begin to invade the wall of the colon, to produce ulcers. This causes diarrhoea

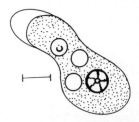

Figure 25.3 A trophozoite of *Entamoeba hystolytica*, with red blood cells that it has ingested. One of these contains a malaria parasite to show how protozoa differ in size. The clear extension is a pseudopodium, an organ of motility. (The bar measures 10 μm.)

which may be mild or sufficiently severe to be called amoebic dysentry, with blood and mucus in the stools. The ulcers can perforate the wall of the bowel, or the amoebae may spread to the liver to cause amoebic hepatitis and abscesses. More rarely abscesses develop in the brain or lungs.

Colonization or infection is diagnosed when cysts or trophozoites of *E. histolytica* are found in faeces, or trophozoites in material taken from the wall of an abscess. The trophozoites survive only briefly when exposed to the air, so collaboration with the laboratory is required if they are to be found. A number of other, non-pathogenic amoebae may also be present in the intestine. Some skill is needed to distinguish between these and *E. histolytica*. Treatment, if necessary, is with metronidazole. Control is by the sanitary disposal of faeces, and the separation of sewage from drinking water and food. The cysts are resistant to chlorine, so chlorinated drinking water is not safe unless its treatment has included efficient filtration.

25.7.2 Other pathogenic amoebae

Some amoebae that live freely in the soil are the causes of rare infections of the central nervous system and the eye. One of them, *Naegleria fowleri*, may be acquired when swimming in contaminated water. It is a very rare cause of an acute progressive meningo-encephalitis.

25.8 OTHER PROTOZOAL INFECTIONS

Trichomonas vaginalis is a common parasite found in the vagina. When numerous their presence is associated with irritation and a profuse discharge. Transmission is predominantly sexual, so it is usually also present in the urethra of the male partner, in whom it rarely causes symptoms. Diagnosis is by microscopy. Treatment, of both partners at the same time, is with metronidazole.

Giardia lamblia may cause an intractable diarrhoea that is more commonly, but by no means exclusively, acquired in tropical and subtropical regions. It multiplies in the duodenum, and may interfere with the absorption of nutrients. It produces cysts that leave the body in the faeces so its mode of spread, diagnosis, treatment and control are as for *E. histolytica*.

Balantidium coli may also cause diarrhoea. It lives in the colon and produces cysts, so is also dealt with in the same way as *E. histolytica*.

Cryptosporidium parvum is a common intracellular parasite of the lining of the intestinal tract of humans and animals. An initial infection, often in childhood, lasts for two or three weeks, with diarrhoea and abdominal pain, rarely with vomiting and fever. Cysts are produced, and infection is often acquired from drinking water. In immunodeficient individuals, particularly in patients with AIDS, the infection is not self-limiting and diarrhoea persists to the extent that life may be threatened. The parasite is very small, so microscopical diagnosis requires special care. The cysts are difficult to kill and treatment is not very effective.

FURTHER READING

1. Hein, T.T. and White, J.F. (1993) Qinghaosu. *Lancet*, **341**, 603–8.
2. See Appendix B.

26 The helminths

(*Ascaris, Toxocara, Enterobius, Trichuris*, Hookworms, *Strongyloides, Trichinella, Filaria, Dracunculus*, Tapeworms, *Schistosoma, Paragonimus, Clonorchis, Fasciolopsis*)

26.1 INTRODUCTION

The numerous parasitic worms that may infest the human race fall into three classes. They are the **nematodes** or roundworms and two varieties of flat worm, the **cestodes** or tape worms, and the **trematodes** or flukes. Worms cause little trouble in the more developed parts of the world, but they contribute to a great deal of ill-health in less developed areas. Many of them have complicated life cycles. They may live in various tissues of the body, or on its surface in the intestine, or they may move between these sites. Some worms complete their life cycles in the same individual and others move between different hosts. In small numbers they may do little or no harm but when numerous they announce their presence in a variety of more or less unpleasant ways. An eosinophil polymorphonuclear leucocytosis is commonly found in the blood of individuals infested by worms (Box 26.1).

Worms are generally visible to the naked eye and, at least in their mature stages, have complex structures. Adult worms measure between a millimetre and several metres in length. Their eggs or immature larvae may be acquired

---

Box 26.1 The worms

Worms have certain features that make them unique among human parasites. One is that adult worms do not multiply in their hosts. This is why some prefer to speak of the presence of worms as an infestation rather than as an infection and why individual infestations most commonly involve a small number of worms. When the numbers are larger the increase tends to happen insidiously over a period, in places where the standards of hygiene are low. The heaviest infestations are therefore found in parts of the world where malnutrition is also common. Here worms cause a great deal of morbidity and may also contribute significantly to avoidable mortality.

directly by the faecal-oral route, or they may be consumed in meat, fish, shellfish or other crustaceans or insects, or on vegetation. Alternatively the larval forms may penetrate the skin either by themselves or with the help of a blood-sucking insect. Some of the infestations they cause are rare, and limited to restricted geographical locations. Those that are more common and widespread are dealt with here, in summary form.

26.2 THE NEMATODES

The nematodes, or roundworms, have a tubular structure, like earthworms, with a mouth, a gut and an anus. Some lay eggs, and others release their young as larvae. The sexes are separate, so an infested individual who produces fertile eggs or larvae must be host to at least two of them. Some are only acquired from other people, and some from animals. Their adult forms are found either in the intestine, or in tissues.

26.2.1 The intestinal nematodes

(a) Ascaris lumbricoides

Human roundworms look like large, smooth, pale earthworms. They live in the gut and at maturity they reach a length of between 150 mm and 300 mm. They are found in all parts of the world though they are more frequent in areas with poor sanitation and particularly where human faeces is used as a fertilizer. Ascaris eggs pass out in the faeces and begin their development in the soil. If they then contaminate food or drink, or if soil is eaten (pica), they enter another human intestine. Here the larvae hatch out and they penetrate the wall of the gut to reach the blood stream and are carried to the lungs. There they escape from the blood into the alveoli and pass up the bronchi and the trachea to the larynx, where they are swallowed. Once more back in the intestine they grow into adults and complete their life cycle (Box 26.2).

The migration of small numbers of ascaris larvae and the presence of a few adults in the gut produce little or no disturbance in most people. Larger

---

Box 26.2 The common roundworm

A female ascaris at the height of her powers may lay 250 000 eggs a day and might live for a year. If human faeces are not disposed of hygienically the soil around houses may contain 100 fertile ascaris eggs in each gram. It is not surprising that perhaps a quarter of the people in the world are hosts to ascaris, with an average of seven worms each. Although most people do not suffer as a result, people who are already malnourished cannot afford to feed large numbers of voracious worms as well as themselves. With so many people infested, more serious illness is estimated to result in about a million people a year, and perhaps 20 000 of them die.

In 1947–48 an experiment was conducted in Germany in which an area of farmland was fertilized with human sewage. The result was an epidemic of ascariasis in the local population.

numbers can provoke an allergic reaction as they migrate through the lungs. In the gut they may cause abdominal discomfort or, rarely, intestinal obstruction. The diagnosis is made if worms are seen in faeces or vomit, or if the characteristic eggs are found in them, by microscopy. Treatment is with drugs that paralyse the worms so they are washed out of the gut by the flow of its contents.

(b) *Toxocara canis* and *Toxocara cati*

The adults of these worms live in the intestines of foxes, dogs and cats. As happens with *Ascaris* their eggs contaminate the soil, so may accidentally reach a human host. Here the larvae hatch and commence their migration, but in what for them is an unnatural host they lose their way. They do not re-enter the gut or achieve maturity, but wander aimlessly about in the tissues. In most cases this causes no symptoms, though there may be an eosinophilia. A few individuals suffer allergic symptoms, with fever. Very rarely the larvae enter the eye and cause blindness. The diagnosis is by serology.

Puppies and kittens acquire infestations from their mothers while in her uterus, or from her milk. Between 2% and 3% of human adults in more developed countries have evidence of an infestation with *Toxocara*; the rate is higher in less-developed areas. Despite their high prevalence it is rare to encounter disease due to them. Exposure to puppies or kittens and to contaminated soil are significant risk factors. The eggs of roundworms in soil (or in children's sandboxes) are not destroyed by disinfectants.

(c) *Enterobius vermicularis*

Children are the usual victims of infestations with threadworms (pinworms, between 5 mm and 10 mm long). Threadworms are the most common human nematode parasites in developed countries. They cause very little disease. Female worms emerge through the anus at night, to lay their eggs on the perianal skin. This may cause an itch, and the victim scratches the area. If the child then sucks its fingers the eggs re-enter the gut and hatch to produce a further crop of *E. vermicularis*. This is why there may be hundreds of worms in a single child. In girls the worms can invade the vagina, to produce irritation and inflammation.

The diagnosis is made if the eggs are found on the perianal skin. A piece of transparent adhesive tape is applied, sticky side to the skin, first thing in the morning. This is then examined under the microscope. Treatment is relatively simple, but reinfestation from others may be a problem in families and schools.

(d) *Trichuris (Trichocephalus) trichiura*

The whipworm is 30 mm at 50 mm long. The anterior two-thirds is thin and threadlike, and with its thicker posterior part it resembles a tiny whip. The thin anterior end is threaded into the wall of the intestine, for attachment. The life cycle of the whipworm, and the ways in which infestations due to it are diagnosed and treated, are similar to *Ascaris*. Infestations rarely cause symptoms, though if hundreds or thousands of worms are involved the individual may suffer from a disease that resembles a chronic form of dysentery.

(e) The hookworms

Adult hookworms are about 10 mm in length. They attach themselves to the lining of the intestine by their mouths, and feed on the mucosa and on blood. The eggs are passed in the faeces. In the soil they develop into larvae that can penetrate the skin, usually of the bare feet. They travel in the blood to the lungs, and then reproduce the final migration of *Ascaris* to reach the intestine. Among poorly nourished people with heavy infestations the loss of blood leads to anaemia and debility (Box 26.3). There are two varieties,

---

Box 26.3 Bloody worms

It has been estimated that each adult ancylostome consumes 0.15 ml of blood every day. For a well-nourished person with a few worms this loss is negligible, but the situation is different for an individual infested with 1000 of them. The loss of blood is now over 100 ml a day. This scale of haemorrhage for months on end would not easily be supported even in the most favourable circumstances. Some anaemia is almost inevitable, and for the malnourished the balance between life and death becomes an issue.

Ancylostoma duodenale and *Necator americanus*, the latter supposedly exclusively limited to the Americas, though both are spread more widely than this suggests. The infestations are diagnosed when microscopy reveals the presence of the characteristic eggs in the faeces. A number of treatments are available, adapted to local circumstances. Other animals have their own hookworms, and as with *Toxocara* spp., their larvae may invade humans to produce accidental, incomplete infestations. The larvae wander in the subcutaneous tissues to produce cutaneous larva migrans, a creeping eruption of the skin.

(f) Strongyloides stercoralis

This worm is widely distributed in the tropics and subtropics. The adults, 2 mm in length, live in the intestine. The eggs hatch inside the gut, and the larvae that emerge follow one of two routes. They may pass out in the faeces to develop into adults that live freely in the soil. These eventually produce more larvae that are able to pierce the skin of a human host to cause a new infestation, in a manner similar to the hookworms. Alternatively the larvae in the gut may complicate and perpetuate an existing infestation if they penetrate the wall of the bowel of the original host. This autoinfestation is normally held in check by the host's immune system, and it happens little if at all in healthy individuals. If the immune system is disordered (as in

AIDS or by immunosuppressive therapy) a hyperinfestation develops with a massive and potentially lethal invasion of the wall of the gut and other organs. The diagnosis of strongyliasis is made when the larvae of *S. sterocoralis* are found in the faeces, or in hyperinfestations in other body fluids. If they are scanty, as happens in long-standing cases in immunocompetent people, serological methods may be used. Treatment is difficult, and the drugs employed are toxic.

(g) Trichinella spiralis

The mature parasite is found in the intestines of carnivorous animals, including rats. Adult females (2–14 mm in length) produce a total of about 1000 larvae each. These penetrate the wall of the gut and migrate to muscle, where they form cysts. When an animal with these cysts is eaten by a predator the larvae are released and develop into adults in the predator's intestine, to complete the life cycle. Domestic animals may be infested if they are fed unheated animal products.

Human infestations are most frequently acquired from undercooked or raw pig meat, or sausages made of it, or 'beefburgers' that contain it. In the Arctic, bears, seals and walruses replace the pig. The disease used to be common in several developed countries, but better animal care and improved food hygiene have reduced the risk. It is still common where appropriate precautions are not applied. The larvae are killed by cooking that makes the meat change colour, and, for non-Arctic species, by freezing.

In the human host the life cycle is the same as it is in animals. Whether or not there are symptoms depends on the number of parasites involved, and host susceptibility. The gastrointestinal phase may be accompanied by nausea, vomiting and diarrhoea. Larvae begin to cause a reaction in the tissues one or two weeks after the infested meal has been consumed. In this phase the individual may suffer from pains in the muscles, a swelling of the eyelids, a variable fever and general debility. In very heavy infestations the heart is involved, perhaps fatally.

Diagnosis may be clinical, or is made by microscopy of a fragment of muscle, or by serolgy. Treatment of an established case can only be supportive, as by the time a diagnosis is made it is usually too late to improve the situation by an attack on the adult worms in the gut. Steroids may help to alleviate the inflammatory and allergic responses to the larvae, when these are severe.

26.2.2 The tissue nematodes

(a) The filarial worms

Of the eight filarial worms that may infest humans, three are of major importance. *Wuchereria bancrofti* is found in tropical areas worldwide, *Brugia malayi* in South East Asia, and *Onchocerca volvulus* in Africa and Central and South America. They are spread by biting insects, mosquitoes, flies or midges. The adult worms live in the lymphatic tissues or beneath the skin.

They release their offspring (called microfilariae) into the blood or the skin, where they wait for their insect vectors to take a meal. In some cases they time their release to coincide with the biting habits of the insect concerned.

Any disease that results develops slowly, with an intensity that depends on the number of parasites present. The pathology is of two kinds. That due to the microfilariae is mainly allergic in character. In the case of *W. bancrofti* and *B. malayi* the reaction to the adult worms interferes with the drainage of lymph to cause recurrent swellings of the genitalia or other parts of the body. In some people these swellings become permanent, to produce elephantiasis.

Adult *O. volvulus* live in nodules under the skin. When the microfilariae are released they can enter the eye where they are a major cause of the preventable loss of sight called river blindness. The name highlights the fact that the insect vector, a blackfly (*Simulium* spp.) breeds in water, and does not move far from it. It has been estimated that more than 30 million people worldwide are infested with this worm, and that two million have defective vision or are blind as a result.

The diagnosis is made when the microfilaria are seen under the microscope in blood collected at night for the two species that have night-biting mosquitoes as vectors, or that emerge from skin snips taken at any time for *O. volvulus*. Treatment with antifilarial drugs is useful in the early stages of the disease when the adult worms are still active, but is of no use later.

(b)   Dracunculus medinensis

The Guinea worm is found in Africa, India and Pakistan. The adults live in the subcutaneous tissues where the female may reach a length of over a metre. A small ulcer, often on the lower part of the host's leg, develops over the worm's anterior end. Through this she discharges up to 3 million larvae whenever the part is immersed in water. In the water the larvae enter a tiny crustacean (cyclops) in which it undergoes further development. If the cyclops is swallowed when water is drunk the worm escapes from the gut to reach the skin and so completes the cycle. Bacterial infection may spread along the lengthy trail of the worm and this can cause serious prolems. The worm may be removed surgically, or by the traditional method of winding it a little at a time day-by-day onto a matchstick, with care to avoid breaking it. Some drugs, metronidazole for example, may be helpful. The disease can be eradicated by the provision of supplies of clean drinking water.

26.3   THE CESTODES

The tapeworms are named for their resemblance to pieces of tape. They do not have a digestive tract, but absorb their nutrients from the contents of the intestines in which they live, through the surfaces of their bodies. At one end of a tapeworm is a small head, equipped with hooks or suckers that anchor it to the wall of the gut. Below this is a neck and the point of active growth from which **proglottids** are produced. These form one at a time and stream out behind the head like a string of postage stamps that get larger as they get older. Each proglottid contains male and female reproductive organs that mature as they move down the tape. By the time they reach the end and are shed, they are full of fertile eggs. In this state they pass out in the faeces to begin the next stage of their development.

The length of a mature tapeworm depends on the number of proglottids and the size of each of them. It varies in different species between a few millimetres (an eighth of an inch) and 10 metres (33 feet). The life cycles of cestodes characteristically involve two or more creatures. In their definitive hosts the worms develop into their adult, egg-producing forms. The eggs hatch to produce larvae in an intermediate host, two of which may operate in succession in some cases. The cycle is completed when an intermediate host that contains larval cysts or later larval forms is eaten by the appropriate definitive host. Exceptionally, eggs may develop into larvae within a definitive host.

Humans are definitive or intermediate hosts, or both, for at least eight different cestodes. The presence of an adult worm in the intestine is rarely noticed until the individual is made aware of it by the appearance of proglottids in the faeces. Larval forms are a different matter, and these may produce symptoms that depend on their anatomical location, their size and number, and any allergic reactions they cause. Most definitive human infestations are the result of eating raw or undercooked meat or fish. Intermediate infestations are acquired by eating food contaminated by animal faeces that contain the eggs of their tapeworms.

Cestode infestations are avoided by high standards of animal care, and good hygiene. They are significantly less common in more developed countries. The associations that can develop between humans and their more important cestode parasites are outlined in Table 26.1. The intestinal forms of infestation are diagnosed when proglottids are found in the faeces. The tissue phases are diagnosed by serology, or at surgery. When necessary the tissue phases are treated surgically, perhaps with the help of drugs, and drugs can be used to expel adult worms from the gut.

Table 26.1 An outline of the associations that develop between humans and their more important tapeworm parasites

Tapeworm	Host 1*	Host 2*	Distribution
Taenia saginata	Humans	Cattle	Worldwide
T. solium	Humans	Pigs, humans[1] etc.	Worldwide
Equinoccus granulosus	Dogs, foxes,	Sheep, etc.	Worldwide, associated
E. multilocularis	rodents, etc.	humans[2]	with sheep
Diphylobothrium latum	Humans	Cyclops and fish	Northern hemisphere
Hymenolepsis nana	Humans	Humans, beetles, etc.	Tropics and sub-tropics

*Host 1, definitive hosts; Host 2, intermediate hosts
Disease due to larvae in human tissues:
[1]cysticercosis
[2]hydatid disease

26.4 THE TREMATODES

The trematodes, or flukes, are unsegmented flatworms, often shaped like a leaf. Their definitive hosts are a variety of vertebrates, including humans. Their intermediate hosts are always snails, sometimes with other creatures in addition. Other than with the schistosomes, both sexes are represented in a single parasite. In humans the adult flukes are found in blood vessels, or in the lung, liver or intestine.

26.4.1 Blood flukes

Adult schistosomes (10 mm to 20 mm) live in the veins in the abdomen, and the females lay eggs that appear either in the faeces (*Schistosoma mansoni* and *S. japonicum*), or in the urine (*S. haematobium*). When water is contaminated with these excreta the life cycle of the parasites is continued in snails. In due course the snails release tiny motile cercariae that are able to penetrate the skin of individuals who bathe or wade in the water. Once inside the body these develop into adult flukes.

When cercariae penetrate the skin they cause small haemorrhages. If they are of non-human species of fluke they go no further, but stay in the skin to cause an irritating rash called cercarial dermatitis or swimmer's itch. Human species cause more generalized symptoms that depend on the number that have invaded. Initial allergy is common, with abdominal discomfort as the worms migrate to their final destination. Symptoms then disappear until the flukes start to lay eggs. As these escape from the body they cause damage to the intestine or bladder to cause diarrhoea with blood and mucus in the stool, or the appearance of blood in the urine. In either case the disease, called schistosomiasis, is chronic and the damage caused is progressive (Box 26.4). Cancer of the bladder may be a late complication of infestations with *S. haematobium*.

Box 26.4 Schistosomiasis

The Great Lakes of Africa and the River Nile may have been the cradle of schistosomiasis. For thousands of years the disease has been endemic in Egypt: eggs of *Schistosoma haematobium* were found in a mummy from about 1000 BC. The fluke was first identified by Theodor Bilharz in 1852. When he was in Africa David Livingstone noticed that from time to time some of his porters passed blood in their urine.

Today the disease is thought to afflict 250 million people. Paradoxically the problem is being compounded by the modern development of water reserves. As a result of the Volta River project in Ghana the rate of infestation in the surrounding population rose from 10% to 100%.

Schistosomiasis due to the three species named (and to some others) is found scattered through tropical Africa and Asia, and in South America. The diagnosis is made when the characteristic eggs, different for the three main species, are found in faeces or urine. Drug treatments are available.

26.4.2 Lung flukes

Paragonimiasis or endemic haemoptysis is caused by *Paragonimus wester-manii*. Some other flukes may also be responsible. The flukes are about 10 mm long and they live in the lungs to cause a chronic disease that resembles tuberculosis. The sputum is stained red or brown with blood, and it contains the eggs of the parasite. The infestation is acquired by eating raw or under-cooked crabmeat that contains late-stage larvae of the fluke. These have been infested in turn by earlier larvae that emerged from eggs expectorated into water and began their development in snails. The disease, common in Asia, is also found in Africa and the Americas.

26.4.3 Liver flukes

There are several varieties of human liver fluke, of which *Clonorchis sinensis* is the most common. The treatment of infestations by the different flukes varies, so accurate diagnosis is important. They live in the bile ducts, and their eggs pass out in the faeces. Their life cycles involve first snails and then fish. Human infestations result from eating raw or undercooked fish, practices that are common in parts of the Far East. The symptoms are upper abdominal discomfort to which may be added any that arise as a result of damage to the liver. The eggs can be found in the faeces by microscopy, and drugs are available for treatment. Liver flukes infest perhaps 20 million people worldwide.

26.4.4 Intestinal flukes

The giant (80 mm or 3-inch) intestinal fluke *Fasciolopsis buski*, together with some others, may inhabit the duodenum. If numbers are present they cause diarrhoea and abdominal pain. The eggs pass in the faeces (where they may be seen under the microscope), and after development in a snail the larvae form cysts on vegetation. The cycle is completed when such things as water chestnuts are eaten. Infestations are common in the Far East.

FURTHER READING

See Appendix B.

The noxious 'insects' 27
(Bugs, lice, fleas, flies, midges, mites)

27.1 INTRODUCTION

Among the major subdivisions of the animal kingdom the phylum *Arthropoda* accommodates those creatures that keep their shape by means of a tough outer skin or hard external plates rather than with an internal skeleton. As causes of human discomfort and disease the most important subdivisions within the arthropoda are the classes that are made up of insects and arachnids. Insects, with six legs and a general tendency to possess wings, include the bugs, lice, fleas, and flies. Among the flies are mosquitoes (anopheline and culicine), blackflies (*Simulium*), sandflies (*Phlebotomus*), and the midges (*Culicoides*) as well as the larger biting and non-biting flies. Adult arachnids usually have eight legs and, generally, no wings. They include the spiders, scorpions, mites and the hard and soft ticks. Crustaceans such as lobsters, crabs, shrimps and water fleas (cyclops), together with the multi-legged centipedes and millipedes and the tongue worms are also arthropods.

Arthropods are a problem to the human race in one or more of three distinct ways. These are the result first of a physical attack, second when they use human tissue as a temporary or permanent home, and third when they transmit organisms that cause disease.

A physical attack may be a bite inflicted by the mouth or some other struc-ture at the front end of an arthropod, or by a sting, from its rear. When the purpose of a bite is to allow the arthropod to feed, the volume of blood lost is rarely of importance. Most of the harm done is by substances that are injected at the same time, and this is even more true of stings. The substances injected may be intended to stop blood clotting when a blood meal is taken, or to act as a venom designed to discourage predators or to immobilize an arthropod's prey. These materials may cause a temporary acute discomfort though the more powerful venoms of scorpions or of some spiders and ticks can cause more severe or even fatal disease. Individuals may also react immunologically to materials injected in this way, and suffer inconvenience due to allergy when bitten or stung repeatedly. In extreme cases this allergy may present as potentially lethal anaphylactic shock as the result of, say, a bee sting in a person who has become highly sensitive to bee venom (Chapter 2).

Some arthropods spend part or all of their lives close to or in human tissues, to cause special problems. Human skin is home to the scabies mite, *Sarcoptes scabei*, the cause of a common infestation to be described later. The sand flea *Tunga penetrans* also spends some time just under the skin, usually of pigs, but also of humans. Fertile females bore into the skin, often of the feet. There they grow to the size of a pea, and when ready they make a small hole through which they discharge their eggs. Myiasis is a blanket term for several human infestations with the larvae of some arthropods. It follows when flies lay their eggs or larvae directly into wounds or ulcers, or on things such as laundry put out to dry from which they can gain access to human tissues. Many kinds of fly are involved, including bluebottles and the common housefly. In the tissues the eggs and larvae develop into maggots (Box 27.1). The larvae of some species of fly are able to penetrate into healthy skin.

Box 27.1 Maggots as surgeons?

Open wounds under plaster casts have sometimes been found to be colonized by the maggots of non-biting flies when the plaster is removed. It has been noted that such wounds are often clean and free from infection. It seems that some maggots at least prefer to eat dead and necrotic tissues, and leave untouched those that are healthy. From this the idea has grown that specially bred maggots should be introduced deliberately into accidental traumatic wounds as a rather unaesthetic alternative, or addition to, surgical debridement.

Some arthropods are concerned with the transfer of microbes or other organisms that are human pathogens. The pathogens concerned may be transferred to a new human host mechanically. This can happen when excreta are left exposed so that within a few seconds flies can walk and feed on faeces and then on food. Equally an arthropod that bites two individuals in quick succession might pick up a microbe from the first, and transfer it to the second in the manner of a 'flying needle'. These routes of transmission are of little or no importance compared with the transfers that are possible when human pathogens are able to multiply in an arthropod that feeds on blood. This provides a highly effective mechanism for the transfer and amplification of communicable diseases because a single infected arthropod can bite and infect several people. A summary list of these important biological associations is given in Table 27.1, with references to the parts of this book where the diseases concerned are discussed.

27.2 PROBLEMS WITH SPECIAL ARTHROPODS

27.2.1 *Sarcoptes scabei*

The scabies mite burrows in the superficial layers of the skin to produce tortuous hairlike channels between fingers and toes, on wrists, in armpits, below the breasts and buttocks, and on the genitals. The mites spread

Table 27.1 Infections and infestations transmitted biologically by arthropods, with the chapters where the diseases concerned are described

Pathogens	Diseases	Arthropods
Bacteria		
Y. pestis (Chapter 13)	Plague	Fleas
F. tularensis (Chapter 13)	Tularaemia	Fleas, ticks, flies
B. recurrentis (Chapter 14)	Relapsing fever	Lice, soft ticks
B. burgdorferi (Chapter 14)	Lyme disease	Hard ticks
R. prowazekii etc. (Chapter 15, Table 15.3)	Typhus, etc.	Lice, fleas, ticks, mites
Viruses		
Arboviruses (Chapter 22)	Many	Mosquitoes, sandflies, ticks, etc.
Protozoa (Chapter 25)		
Trypanosomes	Chagas' disease	Reduviid bugs
	Sleeping sickness	Tsetse flies
Plasmodia	Malaria	Mosquitoes
Leishmania	Leishmaniasis	Sandflies
Helminths (Chapter 26)		
Filariasis	Of the lymphatics	Mosquitoes
	Of the skin	Blackflies

between individuals as a result of intimate contacts that last for some minutes. Infestations are at first symptomless and an individual, unaware of his or her condition, may transmit the mite to others. Significant symptoms begin to appear after a period measured in weeks for someone infested for the first time. They are the result of the development of allergy to the mite, its eggs and its excreta. A rash appears that is more itchy at night. This may involve areas of skin where there are no burrows, and at some distance from them. The minor physical damage done by the mite is much aggravated by scratching and secondary infection. The number of mites present is normally restricted by the host's immunological reaction. If this control fails due to immunodepression the mites proliferate prodigiously. The result is widespread crusted or 'Norwegian' scabies, a highly contagious form of the disease.

The diagnosis is made by seeking the mite, which is just visible to the naked eye. A needle is used to open up a burrow. These can be made more conspicuous if ink is applied to the skin and then rubbed off. If mites are not found (and they may be very scanty) a skin scraping is examined under the microscope and search made for the mites' eggs and faecal droppings.

It must be remembered that areas of skin may seem to be involved though they contain no burrows. A number of effective scabicides are available, and treatment may be needed by other individuals with whom the patient has been in contact. Failure to make the diagnosis may lead to treatment of the rash as if it were eczema. The use of ointments that contain steroids makes matters much worse. To relieve the irritation antihistamines may be taken by mouth at the same time as scabicides are applied. There is no need to disinfest the clothing or bedding of people with scabies, except in the case of the crusted form of the disease, when it is vital.

27.2.2 Other mites

Demodex folliculorum is a curiously shaped mite, less than half a millimetre in length. Most people carry these mites on their faces where they live in the ducts of sebaceous glands and in hair follicles. Although they usually cause no symptoms some of the allergies blamed on makeup may be due to them. Other mites cause sarcoptic mange of dogs and similar conditions in various animals and birds. In poorly maintained buildings or unhygienic surroundings heavy infestations with these mites may spread to involve humans, and cause rashes. Animal and bird mites cannot reproduce themselves in human skin, so the problem disappears when the source has been dealt with.

It is widely recognized that house dust can be the cause of such allergies as asthma and hay fever. (Hay fever is not caused by hay, nor is it associated with fever!) In some cases the house dust mite *Dermatophagoides pteryonyssinus* seems to be responsible. This mite feeds on human and animal dander (skin squames). An average human adult sheds about 300 million squames a day, so houses, and in particular beds, provide a rich source of food for these mites. Their faecal pellets, about the size of grains of pollen, are thought to be the principal allergen. Household cleaning and the making of beds causes them to become airborne so they can be inhaled and trigger the allergy.

27.2.3 Lice

Although they are insects, lice have no wings. The three human varieties are *Pediculus humanus* var *capitis, P. humanus* var *corporis* and *Phthirus pubis*, head, body and pubic lice, respectively.

Head and body lice are almost identical, brown 2–4 mm in length. A female lays her eggs ('nits'), about 300 in all, each of them glued either to a hair for the head louse, or to the clothing for the body louse. A week or more later these hatch and the young lice immediately begin to feed on their hosts' blood. They reach maturity in two to three weeks and they may survive for a total of six to eight weeks. The bites of lice produce a very irritating rash ('vagabonds' disease'), which is often subject to secondary bacterial infection.

Persons of all social backgrounds may be hosts to head lice. They are common in schoolchildren, particularly among girls with long hair. They are transferred by close contact or on such things as shared hats or combs. Body lice are more common among people who live in overcrowded, insanitary conditions who wash and change their clothes infrequently. Close contacts, shared clothing and shared sleeping accommodation provide means of transmission. Body lice only venture onto the skin to feed, otherwise they and their eggs are found in the seams and waistbands of clothing. Body lice not only cause a rash but are also the vectors of some important diseases (Table 27.1). Because body lice are associated with overcrowding and poor hygiene the diseases they spread have a similar distribution. Lice congregate when large groups of people are housed in temporary makeshift accommodation after some natural or human-made disaster (Boxes 15.3 and 27.2).

Box 27.2 Lousy nitpickers

Although their hosts may well feel lousy and spend their time nitpicking, body lice have a comfortable existence. They live in their victims' clothing so are never far from their next meal. They become unhappy if the temperature is not kept at about 20°C. If their host becomes too hot (develops a fever) or too cold (dies) the lice leave to find someone in a more temperate condition. This is why the more deadly louse-borne infections spread so rapidly.

Pubic lice are shorter and broader than their head and body relatives, and their legs are equipped with larger, more powerful claws. Their characteristic shape explains why they are commonly called 'crabs', though an alternative suggestion has been the 'butterflies of love'. The crab louse lives in the hairy parts of the body, except the scalp. They use their claws to cling to hairs, most commonly in the genital region, but also on the thighs and trunk, and even in the beard and eyebrows. Infestations are nearly always sexually acquired, though children may get them from their parents. The female pubic louse attaches her eggs to hairs in the same way as the head louse, but in different parts of the body.

An infestation with lice, or pediculosis, is diagnosed either by the identification of the insects themselves, or of their eggs. A strong hand lens ($\times 10$) or a low-power microscope may be used. A variety of insecticides have been employed in the treatment of pediculosis, and lice in different parts of the world have become resistant to some of them. The nits are not always killed by a single application and a second treatment may be needed a week or so later after any nits that have survived the first application have hatched out. Reinfestation is a common problem unless both members of an intimate couple, or the whole of a family group or community in which transmission is taking place, are treated simultaneously.

Lice and their nits are killed by hot water. Clothes and bedding that contain body lice may be laundered in the hot-wash cycle of a domestic washing machine or commercial laundry at 55°C or more, or they may be dry-cleaned. Combs should be washed carefully in hot water, or immersed in insecticide.

27.2.4 Fleas

Fleas are flightless insects that depend on blood for their survival. They cause irritation when they bite, and are vectors of important diseases (Table 27.1). The many varieties that exist have individual preferences for certain hosts but most will feed on others if driven to it. Dog or cat fleas, for example, will feed on humans in the absence of their normal hosts. Fleas spend most of their time and lay their eggs close to, rather than on, their hosts. They are found on floors and carpets, in birds' nests and animal bedding. The eggs hatch to produce larvae. At first these feed on animal or human debris but after a period of weeks or longer they turn into adults that must feed on blood. If a room or house in which an animal has lived, or that housed a group of people in insanitary circumstances, is re-entered after a period the intruder may be attacked by hundreds of voracious fleas. Vacuum cleaners are the enemy of fleas, though the edges of fitted carpets may escape attention. Insecticides are also useful.

27.2.5 Bed bugs

Cimex spp. are dark brown in colour and about 4 mm long by 3 mm across. They have no wings. They infest human dwellings, where they live in cracks and crevices, behind wallpaper, and in furniture, including the frames of beds (Box 27.3). They emerge to take their blood meals while their human hosts

---

Box 27.3 Bug-destroyer to Her Majesty

In less hygienic times elaborate wooden bedsteads with heavy hangings provided ideal nesting sites for bed bugs, which were a serious cause of discomfort and irritation. They could be dealt with by firms of professional bug hunters. Tiffen and Son advertised themselves as 'Bug-destroyers to Her Majesty' (Queen Victoria). Mr Tiffen ' . . . worked for the upper classes only, for carriage company and such . . . I have noblemen's names on my books . . . I was once at work on Princess Charlotte's own bedstead.'

---

are asleep, and their bites, if numerous, are very troublesome. Heavily infested rooms have a distinctive odour due to an offensive secretion produced by adult bugs. They may be controlled with insecticides, but to eradicate them it may be necessary to block up the cracks and crevices in which they live, and seek out and destroy their nests in furniture.

27.3 THE IDENTIFICATION OF INSECTS AND ARACHNIDS

Mistakes in identification are easily made by the uninitiated. If uncertain, seek expert help. If this is available all that is required is to catch a specimen of the offending arthropod, or collect its eggs, and confine them in a suitable container for transmission. In the absence of an expert the task of identification may be attempted by anyone brave enough who possesses a book about the arthropods of medical importance that has plenty of accurate plates, and directions on how to proceed. A low-power scanning microscope and a few simple chemicals and apparatus will be needed, plus sharp eyes, and patience. When the major groups have become familiar a great deal can be done with a good ×10 hand lens or jeweller's loop.

FURTHER READING

1. See Appendix B, especially item 14, for the insects.

Some Common Infections

PART

7

Common infections $\boxed{28}$

(Infections of the cardiovascular, respiratory, urinary and central nervous systems, and food poisoning)

28.1 INTRODUCTION

An individual may perceive infection as disease (clinical infection) or it may be inapparent (a subclinical infection). In either case the body is reacting to damage that follows a change in the distribution of microbes on it or in it. This may result from an alteration of the microbial balance on a colonized surface because of the overgrowth of part of the existing normal flora, or the arrival and establishment of a new microbe. Alternatively there may be an invasion of a surface normally kept clear of microbes (a 'privileged surface', Chapter 2). Any of these abnormal colonizations or infections may extend to involve other tissues of the body, or these may be opened to direct microbial attack by an injury. Finally damage and a reaction to it may follow the penetration of a microbial toxin.

Superficial invasions may cause infections of the skin, for example with staphylococci, streptococci, or a fungal dermatophyte, or they may produce a gastrointestinal infection such as dysentery. Toxins may act locally as happens in cholera, or generally throughout the body as in diphtheria. Invasions of privileged surfaces are common causes of respiratory or urinary tract infections. Both of these organs have large, very thin surfaces that are richly supplied with blood. Microbes that reach and multiply on these delicate surfaces can easily penetrate them to reach the bloodstream and so cause a bacteraemia or septicaemia.

A fully virulent pathogen can cause an infection in a healthy person if it is introduced in sufficient numbers by the correct route. This does not happen very often, for two reasons. First, it can be difficult for a pathogen to reach the right portal of entry, in sufficient numbers. Second, there are not many major pathogens, and the first infection with each of them usually results in an immunity that prevents reinfection. The situation is entirely different in individuals who are immunodeficient. These people can be infected by small numbers of microbes of low virulence (Chapter 4). In this way they fall easy prey to opportunistic pathogens and even commensals that are present on their own body surfaces, or on people or things about them.

These microbes set up tissue infections that begin at the site of a small break in the surface of the skin or the gut, or of the respiratory or urinary tracts. In the absence of a proper immune reaction these initially minor infections spread locally and then spill over into the bloodstream. The microbes that invade in this way vary according to the flora of the site involved. Many of the people concerned are, or recently were, in hospital, where their normal flora is likely to have been changed. The result is that strains of bacteria resistant to antimicrobial drugs commonly cause major therapeutic problems in immunodeficient patients.

Certain defence mechanisms are more active against some microbes than others, so different types of immune deficiency predispose to particular infections. In general deficiencies in the humoral and phagocytic systems favour pyogenic bacterial infections, while defects of cellular immunity favour the intracellular parasites. These are often viruses, but they include such bacteria as the mycobacteria and listeria, and the intracellular protozoa. Because the activities of the various compartments of the immune system are so closely integrated, however, these broad generalizations may appear to break down in individual cases.

The most common cause of immunodeficiency is malnutrition. This affects all the compartments of the immune system, so in the poorly nourished virtually any kind of infection is not only more easily acquired but is also more severe. Infections may themselves cause immunocompromise. The best example of this is infection with the human immunodeficiency virus (HIV), but it also happens in measles and tuberculosis. In children the combination of measles and malnutrition is particularly dangerous. Immunosuppressive drugs given to prevent the rejection of transplants are specially active against the cellular immune system and steroids act to suppress the inflammatory reaction.

28.2 INFECTIONS OF THE CARDIOVASCULAR SYSTEM (CVS)

28.2.1 Septicaemia

The CVS may be invaded by microbes present anywhere on or in the body. The symptomless presence of microbes in the blood may be described as **bacteraemia**, **viraemia**, **fungaemia** or **parasitaemia** as appropriate, or more specifically by such terms as **meningococcaemia** or **candidaemia**. Bacteraemia may be transient and remain symptomless, or the bacteria may be present in the CVS for longer periods. The prolonged presence of microbes in the blood suggests that they are multiplying somewhere in the lining of the CVS, or in a closely related site, if not in the blood itself. This is usually accompanied by symptoms. When these appear the condition may be called **septicaemia**.

Microbes that invade the CVS may have come from naturally or abnormally colonized surfaces or from the sites of established, perhaps minor, tissue infections. Bacteria are the predominant microbes at these sites. Minor invasions or leaks of bacteria into the blood are very common, even in normal people. They result from such everyday activities as chewing or straining at stool. The leaks are larger and more frequent in individuals with poor dental hygiene or, for example, in someone with an infected colonic diverticulum. Microbes may also leak into the CVS from chest infections, infections of the urinary tract, infected burns or other wounds or indeed any kind of infection. Invasions of the CVS may be iatrogenic when leaks result from such interventions as catheterizations, endoscopies and dental treatment. They may also accompany the medical use of the intravenous and intra-arterial routes for therapy, diagnostic monitoring or intravascular surgery. The abuse of drugs by the intravenous route provides another way in which microbes gain direct access to the circulation.

The normal defences of a healthy CVS are able to deal with sizable invasions of microbes. This ability is compromised by illness, anatomical or physiological abnormality, or by an immune deficiency. Some microbes have even developed ways of circumventing the defence mechanisms, so can invade the blood of apparently normal individuals. Examples are *Salmonella typhi* and *Brucella* spp., together with the microbes that are spread by biting insects (Table 27.1). The ability of some of these microbes to survive and multiply within phagocytic cells protects them from death and provides a route by which they are also spread through the body.

Septicaemia may be acute, subacute or (rarely) chronic. The variation depends on which organism is responsible, and on host factors. The most severe form of septicaemia is septic shock. In this case a bacterial invasion is accompanied by the presence of toxins such as the endotoxins of Gram-negative bacteria to which the body's defences over-react and so complicate the picture (Chapter 2). Acute septicaemia is a life-threatening condition, with a heavy mortality (up to 30%, and higher in septic shock) despite apparently adequate treatment. The diagnosis of bacterial septicaemia rests on blood cultures, and its treatment on the nature and sensitivity to anti-microbials of the organism responsible, together with general supportive measures. A trial is being made in septic shock of the effect of the injection of antibodies prepared against the endotoxins of Gram-negative bacteria.

28.2.2 Infective endocarditis

Most of the microbes that leak into the CVS are rapidly cleared away. On rare occasions some part of the lining of the vascular system provides a niche where they may settle and begin to multiply. The result is a self-perpetuating septicaemia. The site involved is often the endocardium that lines the heart, or the folds of it that form the cardiac valves. The condition is then called 'infective endocarditis', or bacterial or fungal endocarditis according to the microbe responsible. The disease may be extremely acute or very chronic, or anything between. Despite the existence of a complete range of severity most cases cluster at either end of the spectrum so a useful distinction can be made between those that are 'acute' and those that are 'subacute'. The essential differences between these forms are shown in Table 28.1.

Table 28.1 The differences between acute and subacute infective endocarditis

	Acute	Subacute
Causative microbe	More pathogenic[1]	Less pathogenic[2]
Prior damage to the heart	Not necessary	Nearly always present
Infection elsewhere with the same microbe	Nearly always[3]	Not necessary
Duration untreated	Days/weeks	Months/years
Response to therapy	Less than half recover	Many can be cured

[1] *Staph. aureus* the beta-haemolytic streptococci and pneumococci are among the more common causes of acute endocarditis

[2] Viridans streptococci (*Strep. sanguis*, *mitior* and *mutans*) and faecal streptococci (Group D, *Entero. faecalis* and *Strep. bovis*) are common causes of subacute bacterial endocarditis, the latter in older patients (Table 6.1). Rarer causes are *Staph. epidermidis*, *Haemophilus influenzae*, *Bacteroides* spp., *Coxiella burneti* and some fungi

[3] Other than when they are introduced intravenously as when drugs are abused, or by a medical procedure

Individuals at special risk of subacute infective endocarditis are those whose endocardium is already damaged, particularly when this involves the valves. The damage may be congenital or due to rheumatic fever, syphilis or to degeneration and atheroma in later life. A new way to produce endocardial damage is the insertion of artificial heart valves and other forms of cardiac surgery. Previously undamaged valves are attacked in the acute form of endocarditis (Table 28.1).

In subacute cases the initial pathology is the adherence of bacteria to damaged endocardium. The bacteria are incorporated in a small bloodclot that develops at the site, and this allows them to multiply protected from the normal defence mechanisms. The 'vegetation' that results gradually enlarges as bacteria multiply and more fibrin and pus cells are added. It develops a friable surface from which particles that contain bacteria break off to settle in other parts of the body.

The diagnosis of infective endocarditis depends very much on blood cultures, not only to confirm it but also to make the right choice for therapy. The prognosis in 'culture-negative' endocarditis is less good than in cases where the causative microbe has been identified and its sensitivity to antimicrobials is known. In subacute bacterial endocarditis the causative organisms are often present in the blood in small numbers, so repeated cultures may be needed to ensure success. When cultures are taken it is necessary to be very careful to avoid accidental contamination. Cultures may be negative if antimicrobial therapy has already been started or if the microbes concerned are unusually slow-growing, difficult to isolate or leak into the blood in very small numbers.

The treatment of endocarditis must be vigorous. High doses of an appropriate antimicrobial are required to ensure that the drug penetrates the vegetations to reach the microbes inside. Two or more drugs may be needed in cases of an infection with a more resistant organism. Treatment must continue for several weeks to prevent a relapse. Surgery may be necessary to excise and replace infected and severely damaged valves.

Individuals whose endocardium is damaged require to be protected when they undergo dental extraction, urinary tract or intestinal endoscopy, cardiac surgery or any procedure that might cause a bacteraemia. A suitable prophylactic antimicrobial is given just before the procedure begins. Penicillin by injection or amoxycillin by mouth are often used, supplemented by gentamicin if penicillin-resistant microbes are likely to be involved. Erythromycin is an alternative for patients allergic to penicillin.

28.2.3 Myocarditis and pericarditis

The muscle of the heart may be involved as a minor part of many generalized infections. More rarely, as happens in diphtheria, myocarditis is a dominant feature. *Toxoplasma gondii* and the Coxsackie B viruses can make a direct attack on the myocardium. The pericardial sac may be involved secondarily to myocarditis, and pericarditis sometimes complicates infections of the neighbouring lung or it develops as a part of a septicaemia. *Mycoplasma pneumoniae* or *Mycobacterium tuberculosis* are common causes of pericarditis.

28.3 RESPIRATORY INFECTIONS

A respiratory infection may involve either the upper or lower parts of the respiratory tract, or begin in the upper part and descend into the lower. In health, the lower respiratory tract (below the level of the larynx) is kept substantially free of microbes. This is achieved by mechanical factors comprising the cough reflex and the broncheo-tracheal mucous escalator together with the other mechanisms outlined in Chapter 2. In the upper respiratory tract healthy paranasal sinuses and the middle ear are also kept free of microbes but the mouth, nose and throat carry a heavy normal flora. This includes species of the genera *Staphylococcus*, *Streptococcus* (including pneumococci), *Corynebacteria*, *Haemophilus*, *Neisseria*, *Bacteroides* and *Actinomyces* together with various enterobacteria and fungi. Latent viruses may also be present, and if reactivated they appear in the saliva from time to time. This is particularly true of the herpes group of viruses.

The defence mechanisms of the lower respiratory tract may be inactivated in various ways. The cilia may be congenitally defective, or their action may be inhibited by drugs (including anaesthetics), tobacco smoke, alcohol, or by infections, particularly with viruses and mycoplasmas, or they may be bypassed medically, by intubation. The cough reflex may be inhibited at the same time, and various of the other mechanisms rendered inactive. Microbial invasion of the lower respiratory tract follows, and this often develops into an infection.

The more common infections of the upper respiratory tract include sinusitis, otitis media, the 'sore throat' syndrome, infections of the teeth and gums and acute epiglottitis. These may be caused by the viruses, bacteria or fungi noted in earlier chapters of this book. Each type of infection may be caused by several different microbes. A good example is the 'sore throat syndrome' an acute infection localized in the tonsils and neighbouring parts of the pharynx. A sore throat may be a part of the early stages of generalized viral infections such as chickenpox or measles, or of infectious mononucleosis (glandular fever). The throat is also sore in diphtheria or Vincent's angina. Of the rest about one-third of cases are caused by Group A beta-haemolytic streptococci (*Strep. pyogenes*) or streptococci of other Lancefield's groups, and the remainder by various viruses, *M. pneumoniae* or *H. influenzae*.

Laryngo-tracheo-bronchitis or croup is a progressive descending infection caused by the measles and other respiratory viruses and in particular by the parainfluenza viruses. The respiratory syncytial virus is an important cause of a severe bronchiolitis in infants. *Bordetella pertussis* may also be implicated.

Bronchitis may be a passing phase of any respiratory infection, but when the bronchial tree is permanently damaged the condition becomes chronic. Any of the microbes that attack the respiratory tract may be responsible, but chronic infections commonly involve such bacteria as *H. influenzae*, *Strep. pneumoniae* and *Branhamella catarrhalis*. Chronic infections are punctuated by acute episodes due to viruses that cause further permanent deterioration. When bronchial damage is severe as happens in bronchiectasis or cystic fibrosis *Pseudomonas aeruginosa* and other pseudomonads are important additional pathogens.

Pneumonia is an infection that involves the alveoli. It may spread through the whole of one or more of the major anatomical sections of the lung to produce lobar pneumonia. Alternatively it appears as an 'atypical' pneumonia in the form of single or multiple scattered patches of inflamed lung on one or both sides of the chest. Classical lobar pneumonia is caused by *Strep. pneumoniae* but a pneumonitis with a similar pattern on an X-ray may be caused by other pathogens. When respiratory viruses cause pneumonias these tend to develop in an atypical form, and this is also true of infections due to *M. pneumoniae*, *Chlamydia psittaci*, *C. pneumoniae*, *C. trachomatis*, *Coxiella burnetii* and *Legionella pneumophila*. Pneumonias due to *Ps. pseudomallei*, *Histoplasma capsulatum* and *Coccidioides immitis* are confined to certain parts of the world.

A lung abscess can arise secondarily to a pneumonia of any type, and inhaled foreign bodies, trauma and tumours are additional causes. They may also be due to the arrival in the lung of an infected embolus carried in the blood from elsewhere in the body. Many different pathogens (including anaerobes) may be involved, and mixed infections are common. A pleural effusion may complicate pneumonia, and this becomes an empyema if the infection spreads into the fluid in the pleural cavity. Pulmonary tuberculosis and whooping cough are respiratory diseases with special features (Chapters 9 and 13).

Most upper respiratory infections are easily sampled with a throat swab but their laboratory diagnosis is bedevilled by the heavy normal flora of the nose, mouth and throat. This often includes potential pathogens, so it may be difficult to decide if a microbe found in a culture is significant. Sputum from the lower part of the tract is often collected by expectoration so it too is contaminated with microbes from the oropharynx. Correct interpretation is even more important in these cases, though no less difficult. Post-mortem examinations have shown that misinterpretations of culture results are made easily, and all too often. Even when samples are collected through a laryngoscope or bronchoscope the specimen may be contaminated as these instruments have to pass through the oropharynx. Better specimens may be collected by more invasive methods, but these are often contraindicated in severely ill patients in whom they would be of most use.

The gross appearance of the specimen can help. It is likely to be a waste of time to examine the specimen if it contains too much saliva. A Gram-stained smear will indicate if it contains pus cells, and it may also show the presence of large numbers of a single type of bacterium. Gram-stained films must be interpreted with caution, however, as there is no certainty that the minute portion that can be examined under a microscope is representative of the whole. Special bacteriological media are required for the detection of *C. diphtheriae*, *B. pertussis* and *L. pneumophila*, so if an infection with these pathogens is suspected, the laboratory must be notified.

28.4 INFECTIONS OF THE URINARY TRACT

With the exception of the outer (distal) part of the urethra, the healthy urinary tract is sterile. The distal urethra is colonized by members of the normal flora of the neighbouring skin, which at that site is likely to include some bacteria from the gut.

The term urinary tract infection (UTI) describes bacterial or more rarely fungal or viral colonizations of the urine accompanied by inflammation of some part of the lining of the tract. Urine is an inanimate substance and cannot itself be infected, so the abbreviated term 'urinary infection' is a misnomer. The tract may be invaded from below (an ascending infection from the urethra) or from above (a descending infection from the kidney), or it may arise by extension from neighbouring structure. Any structural or functional abnormality of the tract predisposes to UTI. The abnormality may be congenital (cystic kidneys, strictures or physiological malfunctions), or acquired (pregnancy, prostatic enlargement, post-infective or traumatic strictures, urinary stones, or a neurological defect). Ascending infections are more common than descending, and they are also more common in females because their urethras are shorter. There is an important causative association between these infections and urinary catheterization, and the link becomes stronger if catheters are carelessly or improperly used.

Infections may be limited to the bladder (cystitis) or ascend to involve the renal pelvis and the kidney itself (pyelonephritis). Transient asymptomatic bacteriuria is common, especially in females, and is usually benign, though in pregnancy it may progress to a pyelonephritis and be associated with the premature birth of the infant. Septicaemia may complicate pyelonephritis, and can be precipitated by instrumentation or catheterization of the bladder in the presence of an infection. Because septicaemia has a significant mortality, urinary catheterization is not an entirely safe procedure. It should not be undertaken for trivial reasons.

The outer part of the urethra is not sterile so it is difficult to collect an uncontaminated sample of urine for bacteriological examination. This is why the simple presence of bacteria in urine is not an absolute index of bacterial infection. The problem is compounded by the fact that urine acts as a culture medium that soon converts a small number of contaminants into a large population of them. To overcome this it is usual to culture a 'mid-stream' specimen of urine (MSU).

An MSU is collected into a sterile container after the initial flow has run to waste to wash out as much contamination as possible from the distal part of the urethra. The collection ends before the dribble of the final flow is reached. Prior to collection the prepuce in the male, if present, is retracted and in females the labia are separated. The specimen should reach the laboratory as quickly as possible to limit the effect of the multiplication of contaminants. If delay is inevitable the specimen should be refrigerated or the bacteriostatic agent boric acid is added to it.

Culture is performed by spreading a measured volume of urine onto a suitable medium, in a standard fashion. After incubation for 18–24 hours any bacteria present are counted, identified and, if indicated, an antimicrobial sensitivity test is performed. Interpretation is by a convention that regards a count of less than 10 000 bacteria per millilitre of urine as contamination, and accepts that a count of 100 000 or greater indicates an infection. A count between these figures is equivocal. The basis for these criteria is that in an infection the bacteria enter the urine while it is still within the urinary tract. They continue to multiply in the bladder at an optimum temperature until the urine is voided. In consequence they are present in large numbers. By contrast contaminants are added at the last moment. If the specimen is cultured without delay, or some preservative measure is applied, they have no opportunity to multiply. If specimens are collected and handled properly contaminants can only be present in small numbers.

Specimens of urine that ought to give negative cultures begin to become 'positive' when unpreserved specimens are kept at room temperature for more than an hour. If any factor is introduced that reduces the time urine spends in the bladder (catheterization,' for example), or if bacteria are allowed to multiply after specimens have been passed, then the convention for interpretation is invalidated, and the results are suspect. If a specimen is collected from a catheter (directly, using a syringe and needle and not from the catheter bag) the laboratory should be informed as they may wish to make a numerical allowance. The same is true of urine collected directly from the bladder with a syringe and needle, by suprapubic aspiration.

The presence in urine of an inflammatory exudate supports the diagnosis of UTI. It may consist of an abnormally large number of polymorphonuclear leucocytes (pus cells) and an excess of protein. Although the presence or absence of an exudate provides useful evidence for the existence of inflammation, it is not completely reliable. Infections can exist in the absence of obvious inflammatory exudates and exudates may have a non-infectious cause.

Escherichia coli accounts for more than 90% of uncomplicated cases of UTI. *Staphylococcus saprophyticus* is an important urinary pathogen in young, sexually active women. A brief course of an appropriate antimicrobial should clear these infections. Other bacteria, especially other enterobacteria, are more common in recurrent infections, and their detection should prompt a search for a predisposing factor. Microbes that are very resistant to antimicrobials, such as *Pseudomonas aeruginosa* or *Candida albicans*, may be found. This suggests that the patient has already had several courses of treatment. In these circumstances it is probably that there is some anatomical or physiological abnormality of the renal tract or that the patient has been subjected to long-term catheterization.

28.4.1 Tuberculosis

Tuberculosis of the kidney is accompanied by symptoms and an inflammatory exudate appears in the urine, but routine cultures are negative. Tubercle bacilli usually appear in the urine in small numbers. Cultures should be attempted from large volumes (100 ml) of early-morning specimens of urine, repeated on three consecutive days, and require special media and techniques. The urine is also examined microscopically for acid-fast bacilli (Chapter 9).

28.5 INFECTIONS OF THE CENTRAL NERVOUS SYSTEM

Although brain abscesses, encephalitis and meningitis are separate entities, the close anatomical relationship between the brain and the meninges ensures that there is an overlap between the three conditions. The meninges close to a brain abscess may be involved in the inflammatory process, so meningeal symptoms and signs and changes in the cerebro-spinal fluid (CSF) may be found in addition to any produced directly by the abscess itself. Similarly encephalitis may be accompanied by some inflammation of the overlying meninges, just as meningitis is associated with some inflammation of the brain or spinal cord beneath it.

Apart from these major infectious processes, changes in the CSF or neurological signs may be produced by infected thromboses of the great venous sinuses in the skull. Sepsis in other anatomically related structures, such as the paranasal sinuses, mastoid air cells and dental abscesses may extend secondarily to reach the cranial cavity. Infections may also complicate compound fractures of the skull, surgical operations or the insertion of drains or shunts. Septic emboli, particularly in bacterial endocarditis, may produce small lesions not amounting to brain abscesses.

28.5.1 Meningitis

Meningitis is divided into **aseptic** and **pyogenic** varieties. The distinction depends on the nature of the cellular reaction to inflammation as reflected in the CSF. A marked cellular reaction that consists mainly of polymorpho-nuclear leucocytes (PMN or pus cells) indicates a pyogenic reaction. Such cases tend to be more acute and severe. A response that involves lymphocytes is characteristic of the aseptic form that is often less severe and occasionally more chronic. Changes are sometimes found that are intermediate between the aseptic and pyogenic patterns, particularly in the early stages of an infection.

Cases of aseptic meningitis are most commonly caused by viruses, and they tend to run a short and benign course. The causes have been described in Part 4 of this book. Bacterial and fungal causes include the tubercle bacillus, various leptospires, *Treponema pallidum*, *Borrelia burgdorferi*, *Cryptococcus neoformans* and *Candida albicans*. These infections may have a subacute onset and be more chronic in nature.

In many parts of the world the most common causes of pyogenic meningitis are *Neisseria meningitidis* (mainly in children and adolescents), *Haemophilus influenzae* (in the first five years of life) and *Streptococcus pneumoniae* (at any age, but particularly in children and the elderly (Box 28.1). Enterobacteria (particularly *Escherichia coli*), group B haemolytic streptococci, *Staphylococcus aureus* and *Listeria monocytogenes* are the more common causes of neonatal pyogenic meningitis. *Staph. epidermidis* is found as a cause of pyogenic meningitis associated with the implantation of plastic devices such as reservoirs, shunts or drains.

Box 28.1 Bacterial meningitis

The three most common causes of bacterial meningitis have several other things in common.

In their fully pathogenic state each is equipped with a carbohydrate capsule. These are the principal determinants of their pathogenicity.

They are all common causes of respiratory infections, many or most of them subclinical. These infections are the sources of the bacteria that cause meningitis.

Unencapsulated forms of each of them are quite often found in the normal upper respiratory tract.

Only one capsular type of *H. influenzae* (Hib) regularly causes infections. The meningococcus has five capsular types, and the pneumococcus over 20. Meningitis is a rare complication of the infections caused by these organisms. The single infection needed to immunize against Hib happens early in life, so Hib meningitis is most common in very young children. The meningococcus circulates less often, and with more types to encounter meningitis may be delayed until later childhood or young adulthood. With many more types the pneumococcus continues to cause meningitis into old age.

28.5.2 Other CNS infections

Most infections of the brain itself are due to viruses, and these have already been described. Encephalitis may present on its own, or be a part of a more general illness. Abscesses of the brain are usually caused by bacteria. They may be the result of an extension of an infection from an adjacent site, or reach the brain through the blood. In this case the origin may be an infection of the lung, skin or endocardium. *Staph. aureus* and some of the streptococci are frequently found but other microbes (including anaerobes, fungi and protozoa) may be involved. An exact microbiological diagnosis is made by the examination of pus drawn from the abscess with a syringe and needle. This is an important prelude to successful treatment.

(a) Diagnosis and treatment

The treatment of bacterial and fungal infections of the CNS should be based on a knowledge of the nature and antimicrobial sensitivity of the causative organism or organisms. This is because most of the bacterial causes of these infections vary in their sensitivity to antimicrobials and only some drugs penetrate into the CSF satisfactorily. Many of these infections progress rapidly to produce irreversible neurological damage or death, so there may be little opportunity to correct an initial error.

Culture of CSF or pus from an abscess is preceded by the examination of a Gram-stained smear. If organisms are seen that have the appearance of one of the more common pathogens an informed initial choice of antimicrobial therapy is possible. In competent hands microscopy is more likely to allow a provisional bacteriological diagnosis to be made than are tests that seek to detect microbial antigens in the CSF. A Gram-stained smear also reveals any cells (lymphocytes or PMN) that are present, so can assist with the diagnosis. Prior antimicrobial therapy is likely to make both the microscopical and cultural examinations negative or unhelpful. In this case a test for microbial antigen may be useful.

Other examinations of CSF that may assist with the diagnosis of an infection are tests for protein, glucose and perhaps lactate. An examination of the white cells in the blood may help in the diagnosis of bacterial infections, and blood cultures may also be positive.

28.6 FOOD POISONING

Food poisoning may also be described as gastroenteritis, enteritis or enterocolitis, depending on which parts of the intestinal canal are involved. The disease has an abrupt onset with vomiting, diarrhoea and abdominal pain. Cases of it are often recognized in outbreaks, and there may be a clear association with the recent consumption of a particular food. The concept of food poisoning is useful, but difficult to define. It covers all kinds of intoxication arising from the recent consumption of contaminated food or drink, whether the toxins concerned are of microbial origin or are poisons that are present naturally or have been added artificially (Box 28.2). The word 'recent' is included to exclude such food-related diseases as typhoid, listeriosis and infestations with the protozoa. Although cholera and shigellosis have short incubation periods and are also associated with food these diseases are not considered as examples of food poisoning.

Food poisoning is a growing problem. Some of the increase is related to the industrialization and mass production, transportation and storage of much of what we eat and drink. The condition is common in hot-climate countries and in temperate areas is seen more often in the warmer part of the year. In countries with poor standards of hygiene not only is it more commmon but it is also more likely to be severe. This is because the main weight of infection falls on the very young who are intolerant of the dehydration it causes and in whom malnutrition may be an added problem. The accurate diagnosis,

---

Box 28.2 Food poisoning

The appearance, smell and taste of food is no guide to its wholesomeness. Some foods are inherently poisonous. Potatoes that have grown above ground contain an excess of solanine; bitter cassava and some fish are toxic all the time, others, seasonally. Some plants and seeds or bulbs and some immature fruits and roots are toxic.

Food may be made toxic due to the use of insecticides, fungicides or herbicides in their cultivation, or other chemicals added during storage or preparation in the kitchen.

Food may become dangerous when grown in soil manured with faeces, particularly if of human origin.

Meat may be contaminated if the animals from which it came was suffering from infection, or by cross contamination from another such animal in the abattoir, butcher's shop or a refrigerator. Contamination may come from an infected food handler, but this is not common other than with viruses.

Food becomes dangerous when small numbers of potential pathogens multiply to reach toxic doses as a result of poor hygiene or inadequate refrigeration.

Food that is mass-produced and widely distributed may cause illness in people who are so scattered as to make it difficult to identify that an outbreak exists.

---

epidemiological study and prevention of food poisoning depend on the availability of competent microbiology laboratories.

The length of the incubation period between the consumption of an incriminated food and the onset of symptoms allows a conveient subdivision of the causes of food poisoning. A short incubation period (five minutes to six hours) is associated with microbial food poisoning due to *Staph. aureus* and the toxins of *Bacillus* spp. that cause vomiting. It is also seen in non-microbial food poisoning that follows the consumption of uncooked red beans, solanine from green potatoes, heavy metals (zinc, arsenic, etc.), and some of the toxins found in fish and mushrooms.

Short incubation period food poisoning tends to be toxic in origin, and nausea and vomiting are the predominant symptoms. The enterotoxins of *Staph. aureus* and *Bacillus* spp. pre-formed in food are not destroyed by brief boiling. Red beans are only toxic if eaten raw or nearly so; the toxin is destroyed if they are fully cooked. Scombrotoxin is a histamine-like substance formed in the flesh of scombroid fish (tuna, mackerel, bonito, etc.) by the action of enzymes and bacteria. The paralytic neurotoxin of dinoflagellates may reach seriously toxic levels in the flesh of filter-feeding shellfish such as mussels and cockles. This happens when these forms of plankton multiply excessively in the sea where they may be visible as a 'red tide'.

In food poisoning a medium incubation period ranges from six to 48 hours. This covers disease due to salmonellas, *E. coli*, *Clostridium perfringens*, *C. botulinum*, *Vibrio parahaemolyticus*, *Bacillus* spp. (diarrhoeal type) and some mushrooms.

Salmonellas and the vibrios cause gastroenteritis by invading the wall of the gut, the others when they produce toxins that are either pre-formed in food (*C. botulinum*) or may also be made within the gut (*C. perfringens*, *B. cereus*). Salmonellas may (rarely) go on to invade the rest of the body to produce septicaemia, meningitis, osteomyelitis or abscesses. *E. coli* can cause enterocolitis by a number of different mechanisms. The disease may present as infantile or travellers' diarrhoea. Cases of food poisoning with a medium incubation period usually present with diarrhoea and abdominal pain. Nausea may be noted, but vomiting is less marked. In botulism there is usually vomiting but no diarrhoea before the neurological symptoms appear. Any botulinus toxin present in food or water can be destroyed by heat (boiling for 20 minutes).

Long incubation period food poisoning may have an incubation period of up to five days, though two or three is more usual. This group includes disease caused by the campylobacters, *Yersinia enterocolitica*, sometimes by *E. coli* and by the gastroenteritis viruses.

In some parts of the world campylobacters are very common causes of gastroenteritis. Most cases are sporadic, and their epidemiology is not clear. Epidemics have sometimes been traced to milk or water, rarely to poultry. Long incubation period food poisoning is characteristically associated with diarrhoea and lower abdominal pain. Nausea and vomiting are usually inconspicuous.

(a) Diagnosis, treatment and control

The various kinds of food poisoning can only be separated fully by laboratory tests. These involve the collection of samples of stool or vomit from those affected, and if at all possible, any residues of the food that might be the cause. Stool examinations begin with microscopy. The presence of pus cells indicates invasive disease, and an excess of red cells suggest a dysenteric picture. Appropriate culture methods allow the isolation and identification of bacterial pathogens and other techniques are used for this characterization of viruses. In some cases epidemiological study requires detailed identification of the causative organism. In an apparent epidemic of salmonellosis, for example, it is necessary to ensure that all the isolations from various patients and the postulated source are indistinguishable. This is because salmonella infections and the finding of salmonellas in food are both common events. It is not sufficient to say that unspecified salmonellas have been found in patients and in food, and then to assume that they must be related. This is particularly important if there is a possibility of legal action against a food manufacturer or caterer.

In most cases of food poisoning the only treatment necessary is to ensure an adequate intake of fluids and any modification of diet that may be imposed by nausea. In severe cases oral rehydration salts (Chapter 11 and Box 11.4) may be used, particularly in small children, or intravenous infusion may be necessary. Antidiarrhoeal drugs are best avoided, as, if they work, they prevent the body ridding itself of the toxic agent. Antimicrobials are of no use in cases in which a toxin is responsible. In other uncomplicated cases it has been shown that their use prolongs the period for which the causative microbe is found in the stool, and that they do not shorten the course of the illness.

An infected individual should not prepare food for others while the diarrhoea persists, and this is particularly important for professional food handlers, from whom negative stool cultures may be required before they return to work. Health care workers should not attend to patients while they still have symptoms.

Many foods contain small numbers of potentially pathogenic bacteria. These tend to multiply best at temperatures between 35°C and 40°C. If they are not to develop into an infectious dose (of the order of 100 000 per gram), food that is to be stored for a short time should be kept hot (over 60°C), otherwise it should be kept cold (below 4°C). For longer periods, food may be stored dry or frozen; a high concentration of salt may be added to it, or it can be made acid. Food is also safe if it is completely sealed in cans or other impervious containers and has then been sterilized by heat or irradiation.

The foodstuffs most often implicated in cases of food poisoning are derived from animals or poultry, including milk and eggs, and any products that are made from, or contain them. Grains and spices are often contaminated with spores of *Bacillus* spp.

The heat applied in thorough cooking kills most pathogenic microbes, though some spores can resist the action of boiling water. Too much heat makes food unpalatable, and in some cases (with shellfish in particular) the line between what is thought of as attractively edible and what is safe disappears. Large joints of meat and poultry require careful cooking to ensure that the heat penetrates throughout. This is impossible if frozen food is not completely defrosted before it is cooked. Careful cooking and efficient refrigeration are the keys to safe food, and cooked and uncooked foods should be kept apart to prevent the recontamination of what has been made safe by cooking. For the same reason work surfaces, mechanical slicers and other utensils need special attention. They should be kept clean and dry.

Food handlers are not often the source of the microbes that cause food poisoning, other than in the cases of staphylococci and the viruses. When a salmonella is found in the stool of a food handler associated with an outbreak of salmonellosis the individual may be identified as the source. This is rarely so. Food handlers sample the food they prepare, so more often than not they are just victims of the outbreak. Although they may be innocent in this respect, food handlers are often to blame for an incident when by ignorance or carelessness they allow wholesome food to become poisonous. Both professional and domestic caterers require education in food hygiene, and many kitchens need to be improved. This is particularly true where large numbers of meals are prepared in crowded conditions. Better animal care can reduce the contamination of food at source. Early and efficient laboratory and epidemiological investigation of outbreaks helps towards the introduction of sensible, effective control measures.

28.6.1 The investigation of an outbreak of food poisoning

The steps taken in an investigation of an outbreak of food poisoning vary with the circumstances, but the history of a simple example is given to illustrate what they might be.

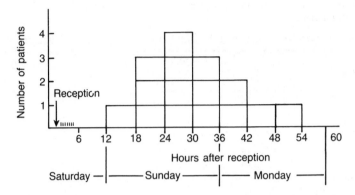

Figure 28.1 The epidemic curve of an outbreak of salmonella food poisoning. Each square represents one patient, placed along the baseline of the histogram according to the number of hours after the reception when their illness started.

Between them, five general practices in a small town reported the appearance of 16 cases of food poisoning in a single week. One elderly lady was admitted to hospital with severe gastroenteritis, where she died. The average monthly incidence of reports of gastroenteritis in the same community was between zero and three. Salmonellas that were indistinguishable were isolated from the faeces of four of the six cases in which culture was attempted. The local Public Health Department investigated all the cases reported to them and found that 15 of the 16 had attended a wedding reception on the previous Saturday. They drew a graph of the time of onset for each of the 15 cases to produce an 'epidemic curve' (Figure 28.1). This histogram revealed that the incubation periods measured from when the reception was held were all consistent with a common-source outbreak of salmonella food poisoning, acquired at that time.

The guest list and the menu of food served at the reception were obtained. Fifteen other guests were interviewed, three of whom reported mild symptoms of gastroenteritis for which they had not sought medical help, but with onsets also consistent with infection at the reception. The others had remained well. There were now 18 cases of apparent food poisoning among a total of 30 guests who had been interviewed. A chart of the food eaten by each of the 30 was prepared (Table. 28.2, A).

The chart revealed that the potato salad was almost certainly the cause of the outbreak as those who had, or had not, eaten it displayed the largest difference in the incidence of symptomms (Table 28.2, B). This is obvious by inspection, but in more complex cases it may be necessary to perform statistical analyses to establish the most likely cause. The fact that two people who had not eaten the potato salad fell ill, and three who had did not, is a common finding in episodes of this sort. It is accounted for by faulty memory or by the fact that diarrhoea and vomiting may be due to a variety of causes (too much to drink at a wedding reception, for example) or that some who helped themselves to the potato salad left most on their plate and ate too little to consume an infectious dose.

Table 28.2 The food consumed by guests at a wedding reception, and its association with gastroenteritis

A) Types of food consumed, and the number of guests with and without symptoms.

Menu item	With symptoms		No symptoms	
	+	−	+	−
Starter	14	4	10	2
Salmon	17	1	11	1
Green salad	15	3	11	1
Mayonnaise	14	4	9	3
Potato salad	**16**	2	**3**	**9**
Fruit salad	13	5	12	0

+, ate, or −, did not eat, this item

B) Proportional differences in the association of symptoms with the types of food consumed.

Menu item	Illness % in those who	
	ate	did not eat
Starter	58	67
Salmon	61	50
Green salad	58	75
Mayonnaise	61	57
Potato salad	**84**	**18**
Fruit salad	52	100

When the caterer was questioned it emerged that the mayonnaise was made with raw eggs, but that it had been refrigerated. The potato salad had been prepared about five hours before the reception. The mayonnaise was added to the potato, but the mixture was kept at room temperature because the refrigerator was full. The outbreak was due to salmonellas from the eggs used to make the mayonnaise. These multiplied rapidly in the potato salad in a warm kitchen, but they were held safely in check when the mayonnaise was stored in a refrigerator.

FURTHER READING

See Appendix B.

Dealing with Infections

The diagnosis of infections 29

29.1 DIAGNOSIS

29.1.1 Clinical diagnosis

Most infections are self-limiting and many are trivial. It makes little difference if a mild infection is diagnosed, misdiagnosed or remains undiagnosed. The outcome is the same whatever treatment is applied, and indeed if drugs are used some people will suffer more harm from their side-effects than from the infections themselves. Severe infections are quite a different matter because a correct diagnosis is a critical prelude to successful treatment, and there may be little time for a second guess. In extreme circumstances the difference between life and death depends upon a response that is both quick and accurate.

A diagnosis is made after a history of the illness has been taken, and the patient has been examined. Classical cases of chickenpox, measles or mumps may be recognized across a room, though as a result of vaccination many young health professionals have never seen cases of the latter infections. Experienced practitioners will diagnosis many other infections in a few minutes. More difficult cases take longer, and may involve the performance of laboratory tests. These are requested to assist with a choice between a number of possible clinical diagnoses, or when a firm diagnosis has been made, to select an appropriate antimicrobial for use in therapy. Some tests are relatively simple and may be done in a small annex or side-room in a clinic or ward. Others are more complex and are performed by specialist staff in fully-equipped laboratories.

The history of an illness can suggest that it has been caused by an infection. Any combination of symptoms that include the sudden onset of a fever with chills, shivering, sweating, pains in the muscles, a rash, a dislike of bright lights, vomiting, diarrhoea and a distaste for food may lead to this conclusion. Other complaints may point to the type or location of an infection. Tightness or pain in the chest when breathing accompanied by a cough suggest an infection in the lower respiratory tract, while an urge to pass water frequently and discomfort when this happens is evidence of an infection of the urinary tract. Enquiry about recent travel may provide an important clue to diagnosis in obscure cases.

An examination may confirm the existence of a fever, and if the temperature is observed over a period a pattern may be noted that suggests a particular kind of infection. The presence of a rash, its type and distribution, or of enlarged lymph nodes or the discovery of a swollen or tender organ are signs of vital diagnostic importance. An X-ray or an examination by one of the newer imaging techniques may be very helpful at this stage.

It is important to note that any or all of the symptoms and signs mentioned may be found in a patient with an illness that is not due to an infection. Certain malignancies, particularly those that involve lymphatic tissues, some forms of arthritis and such diseases as sarcoidosis may be accompanied by a fever. Some patients use considerable ingenuity to fake a fever for reasons that may be conscious or subconscious. It is equally important to note that a serious infection can exist in the absence of fever or any of the other more usual diagnostic features. This is particularly true in patients whose immune responses are less than optimal.

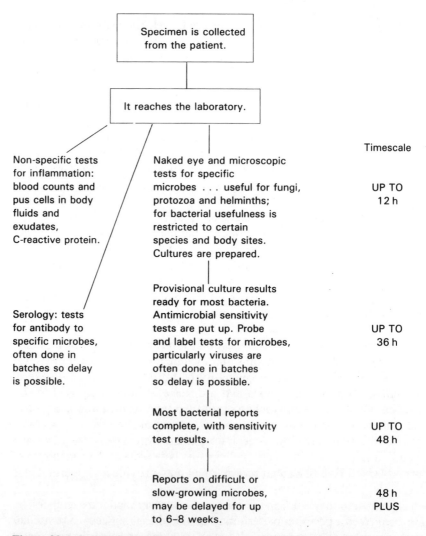

Specimen is collected from the patient.

It reaches the laboratory.

		Timescale
Non-specific tests for inflammation: blood counts and pus cells in body fluids and exudates, C-reactive protein.	Naked eye and microscopic tests for specific microbes . . . useful for fungi, protozoa and helminths; for bacterial usefulness is restricted to certain species and body sites. Cultures are prepared.	UP TO 12 h
	Provisional culture results ready for most bacteria. Antimicrobial sensitivity tests are put up. Probe and label tests for microbes, particularly viruses are often done in batches so delay is possible.	UP TO 36 h
Serology: tests for antibody to specific microbes, often done in batches so delay is possible.	Most bacterial reports complete, with sensitivity test results.	UP TO 48 h
	Reports on difficult or slow-growing microbes, may be delayed for up to 6–8 weeks.	48 h PLUS

Figure 29.1 Outline flow chart for the diagnosis of infections from specimens after they have arrived in a laboratory.

29.1.2 Laboratory diagnosis

Laboratory tests can assist with the diagnosis of an infection in two ways. Some give a non-specific indication that a particular illness is due to an infection and perhaps point to the general class of microbe responsible. Other tests demonstrate the presence of a particular pathogenic microbe so lead to a more precise and specific diagnosis. In either case the first step is to collect an appropriate specimen or specimens, and transmit them to the laboratory. An outline of what may happen to them there, and how quickly, is given in Figure 29.1.

29.2 SPECIMEN COLLECTION AND TRANSMISSION

The specimen or specimens to be examined are selected in the light of the clinical diagnosis that has been made. The possibilities have been described in Parts 3–7 of this book. It is normal practice for a laboratory (or a stores organization acting for it) to supply the containers in which specimens are transported, and some of the apparatus (swabs, spatulas, spoons and so on) that may be used to collect them. Request forms that must accompany the specimen usually come from the same source. Laboratories normally provide instructions on when and how the containers are to be used, and how to complete the forms. Completed request forms must identify the patient, indicate the nature of the specimen, name the test required and give any other information asked for to assist the laboratory to respond in a helpful and efficient way. A busy laboratory may receive several hundred specimens a day. It is important to take some care with the documentation, and make sure that the container is also labelled so that the specimen can be matched with the right request form when they reach the laboratory (Box 29.1).

Box 29.1 Laboratory request forms

The importance of the accurate documentation of laboratory requests ought to be self-evident, but in practice forms are all too often completed carelessly. When the forms that arrived in the microbiology laboratory of a teaching hospital were checked fewer than 10% were found to have been completed correctly in all respects. The part of the form that asked about antimicrobial therapy was left blank in 82% of cases, though when urine specimens were tested for the presence of antimicrobial drugs, 33% were found to contain them. Only 5% of the accompanying forms noted that the patient had been given one of them. The clinical state of the patient was not mentioned on 28% of forms, and the date and time the specimen was collected was left blank on 21% and 83%, respectively. There was good evidence that even when a date was given this more often related to the completion of the form than to the collection of the specimen, which may have been as much as 24 hours later. It should be possible to do better than this.

Specimen containers may be made of metal, glass or plastic, and be re-usable or disposable. They are often sterile, and specimens for microbiological examination in particular should be put into them carefully to avoid extraneous contamination that might confuse the result. Some containers may be provided with something already inside them, such as a swab, or a spoon to collect faeces, or they may contain a powder, liquid or jelly. These are provided as appropriate to prevent clotting when liquid blood is required for a test, or as a fixative for tissue samples that are to be examined by microscope only, or as one of the transport media that are used to keep bacteria or viruses viable but in a suspended state of animation (so far as possible) prior to examination. Special containers are used for blood cultures. If a specimen is put into the wrong one it may be spoiled.

Laboratories take care to provide containers that will allow specimens to reach them in the best possible condition. Circumstances vary and so do the containers supplied by different laboratories. Account is taken of any local safety regulations that apply to this activity, the distance specimens must travel, the time this takes, laboratory patterns of work and techniques, and of course, how much money can be spent. The quality of the work done in a laboratory cannot compensate for inadequate specimens. Care taken in the collection of specimens, the completion of request forms and the speedy transmission of both will be rewarded, and patients deserve no less. It is certainly not a waste of time to study the handbook or other information about the collection and transmission of specimens provided by the laboratory to which they are to be sent.

Some specimens are dangerous because if they are handled carelessly they might transmit an infection to those exposed to them. Good quality containers are designed to minimize the risk of this, but if the cap is not secured properly or if the bottle or tube is dropped, the specimen may leak. Many countries have established national codes of practice that regulate the way clinical specimens are collected, packaged, transported and examined, and what to do in the event of an accident. In other cases individual hospitals have devised their own sets of rules.

Specimens from patients with certain infections thought to be unusually dangerous may be designated and treated as particularly hazardous. The more stringent regulations and heightened precautions that then apply are designed to protect against the worst imaginable accident, so tend to give an exaggerated idea of the real level of risk. Health care workers ought (and in some places must by law) make themselves familiar with the health and safety regulations that apply to them. It is the responsibility of managers to provide safe working conditions and to see that rules are obeyed. They should also ensure that members of staff are familiar with the dangers to which they might be exposed and that they understand the relevance of the measures designed to protect them.

29.3 NON-SPECIFIC TESTS FOR INFECTION

Infections and infestations stimulate the body's immune system. This causes a change in the number of cells responsible for phagocytosis or for the production of a humoral immune response (Chapter 2). Changes in the tissues are usually reflected in the circulation and these are revealed by an examination of the blood. Blood is easily sampled and examined. A change in the number of white cells in the blood is revealed by a 'total white-blood-cell count'. The presence of more white cells than usual is called a **leucocytosis**. The various kinds of white cell that circulate in the blood can be distinguished and counted separately to give a 'differential white-cell count'. The result may indicate the presence of an infection, and give a clue to its nature. An excess of neutrophil polymorphonuclear leucocytes (pus cells) is often seen in pyogenic bacterial infection. An increase in the number of lymphocytes suggests a virus infection, and eosinophils are found in worm infestations. These generalizations have important exceptions that may themselves be diagnostically helpful, for example in enteric fever, whooping cough and glandular fever. Another test that indicates the presence of an infection is an increase in the level of C-reactive protein in the blood (Chapter 2).

The urine of feverish patients often contains a small excess of protein. If more than a trace is present together with numerous pus cells the infection may be located in the urinary tract itself. When numbers of pus cells are found in sputum, pleural fluid, faeces or an exudate from some other part of the body the cause is likely to be a bacterial infection at or near the site. The diagnostic importance of the presence of pus cells or lymphocytes in the cerebrospinal fluid was discussed in Chapter 28.

Some or all of the tests for these general signs of infection may be performed by the staff in clinics or hospital wards. This has the considerable advantage that the results are available at once. The disadvantage is that the tests may be carried out by individuals with little or no training in the techniques involved, who perform them infrequently, with no understanding of how they work or of the pitfalls or dangers inherent in them. No quality control is applied and the reagents used may have been stored wrongly or are out of date. Those who have tests done in these conditions need to be clear about the reliability of the results. If specimens are sent to the laboratory, these non-specific tests are likely to be carried out by properly trained staff who perform them regularly, use reliable reagents and are subject to some form of quality control.

It is important to remember that the results of these non-specific tests do no more than indicate the presence of an inflammatory reaction. Although this may be caused by an infection it can also result from physical or chemical injury.

29.4 SPECIFIC TESTS FOR INFECTIONS

These tests may be direct or indirect. The direct approach requires the examination of a specimen taken from a presumed site of infection in an attempt to identify the causative microbe. In an indirect approach a specimen of blood serum is tested to detect and quantify antibodies that have developed in response to an infection with a particular organism.

29.4.1 Direct tests

Some parasites can be identified with the naked eye, or with a simple magnifying glass. This applies to many insects and most worms. Many protozoa and fungi are identified exactly, or sufficiently for most purposes, by ordinary microscopy. The electron microscope can be used to detect viruses, though it cannot distinguish between those that have the same structure but cause different diseases (Chapter 16). The presence of bacteria can be confirmed by light microscopy, though for most purposes this gives insufficient information (Chapter 5). A positive finding may be of value in some cases, but a negative result only indicates that any microbes that might be responsible are present in such small numbers that they cannot be seen.

Box 29.2 Microscopy versus culture

Direct optical microscopical examination of specimens will be negative when the number of bacteria they contain falls below 10 000/ml. Ordinary methods of culture will detect the presence of between 100 and 1000 bacteria/ml, while enrichment techniques can pick up a single bacterium in 100 ml.

The presence of much smaller numbers of microbes can be detected by culture (Box 29.2). This involves the provision of conditions in which they can multiply in the laboratory. The conditions vary according to the nature of the expected pathogen, as indicated in Chapters 5 for bacteria, 16 for viruses and 24 for fungi. When enough time has elapsed to allow sufficient multiplication, likely pathogens may have to be separated from any other microbes present at the same site. When this has been accomplished any of interest that appear are identified. This nearly always involves microscopy of material taken from the culture and in certain circumstances this may be all that is required. More often further cultures are made under a variety of conditions designed to force the microbe to reveal its identity.

Bacterial pathogens are dealt with by culture more often than the other groups of microbes. This is because, in the laboratory, most of them grow easily and relatively quickly. Cultures are also useful because bacteria are particularly likely to develop resistance to the drugs used to treat the infections they cause. A major part of the work of bacteriology laboratories is the performance of 'sensitivity tests'. These are designed to test pathogenic bacteria isolated from specimens against appropriate ranges of antimicrobial drugs. These tests are done in a variety of ways, but the results need to be interpreted according to standard criteria, under strict quality control. This is to ensure that reports of sensitivity or resistance to antimicrobials have some meaning when treatment is attempted, and that results are comparable between laboratories.

Another way in which microbes may be identified is to locate a unique molecule somewhere on or in it. A test is devised that can recognize the presence of this molecule in a specimen by the use of a probe that seeks out the selected molecule and binds firmly to it. This happens in virus neutralization tests (Chapter 16). In this case the probe is an antibody that can only unite with its own virus. The union between the two is detected because the now-neutralized virus cannot infect an animal or a cell culture that is normally sensitive to it. This approach has been modified to allow a report to be available in a matter of minutes rather than the days required when culture methods are used. The modification involves the attachment of a label to the probe molecule so that its presence is signalled in an unmistakable way.

These **'probe and label'** techniques have many very practical applications in the rapid diagnosis of infections due to viruses and other microbes. The 'probe' must unite firmly with some part of the microbe it is designed to seek, and hopefully with nothing else. The union between the probe and its microbial receptor must be firm enough to withstand the washing process that removes any superfluous reagent, so only probe that has been bound to the microbe is left behind, to show up later. Before it is used the probe has been labelled in some way that allows it to be located when a search is made for it (Box 29.3).

Box 29.3 'Probe and label' techniques

(Designed to locate, identify and quantify antibodies or microbes in specimens from patients.)

Preliminary Steps:
Identify a unique molecular site ('receptor'), part of the item to be detected.

Prepare a probe[1] that will unite firmly with this receptor, and nothing else. Label [2] this probe to make it easy to detect.

The test:
Apply probe to the specimen. If receptors are present the probe will bind to them. Wash off any surplus unattached probe. Seek the label[2] to detect, locate and quantify the receptor, and so the microbe.

[1] The probe may be an antibody to the receptor, an antigen, or homologous RNA or DNA, etc.
[2] The label, and its detection: a dye that converts uV to visible light, or a radioactive compound plus a radiation detector, or an enzyme plus a substrate that changes colour in its presence, or latex particles that aggregate into clumps (plus other strategies).

The label used in fluorescence microscopy is a dye that converts uV to visible light (fluorescence antibody test, FAT). Several other labels are employed. They may be microscopic spheres of latex that congregate into visible clumps when the probe molecule they are attached to reacts with its specific receptor (latex agglutination or coagglutination tests). The label may be radioactive, to be located later with an X-ray film or some more sophisticated radiation detector (radioimmunossay, RIA). In some cases the label is an enzyme that is detected, usually by a change in colour, when a specific substrate that is attacked by the enzyme is added (enzyme-linked immunosorbent assay, ELISA). Probes are of different kinds. They may be antibodies to microbial antigens, or fragments of nucleic acid that unite with equivalent or complementary fragments present in the microbe for which they have been prepared.

When microbial antigens are to be detected these techniques depend on there being sufficient receptors present to which identifiable amounts of the probe can bind. This is not always the case. One way in which this difficulty is overcome is to culture the specimen briefly so the microbe multiplies until sufficient receptors have been formed. Another approach is to apply a relatively new technique called the polymerase chain reaction (PCR).

In theory the PCR can recognize the presence of a single microbe, or even a part of it, in a specimen. In practice a few hundred may be required. A probe recognizes the presence of the microbial receptor for which it is specific, and the polymerase part of the reaction converts this receptor into several million copies of itself. They are now easy to find and the microbe can be identified by one of the other techniques that have been mentioned. A penalty for this sensitivity is that the receptor recognized and amplified may not be the cause of the problem under investigation, or even may not have come from the specimen under examination. At these sub-microscopic levels specimens are easily contaminated. The contamination may come from the laboratory itself, where millions of copies of the same receptor have recently been made from a different specimen.

There is a general difficulty with all probe and label methods when they are applied to specimens taken directly from patients. A positive result does not indicate if the microbe detected is alive or dead, or if a virus is active or inactive. The detection of a small group of dead microbes (or inactive viruses) may not indicate that the patient from whom the specimen came is, or even was, suffering from an infection with that microbe. For example a specimen taken from a patient who has been successfully treated for a chlamydial infection will continue to react positively in tests for chlamydias until all the dead remnants have been cleared away.

These tests are only available if someone has taken the trouble to identify, prepare and label the probe molecules that are specific for the microbes it is desired to detect. As time passes the number of these reagents grows, but by no means all known pathogens are catered for. In addition any pathogens not yet discovered will remain hidden forever if these methods were to be used on their own. Another disadvantage is that probes able to determine the sensitivity or resistances of microbes to antimicrobials are not yet available, so these tests still depend on cultural techniques.

Probe and label tests are technically demanding to perform, and the reagents used are expensive and not everywhere available. Each set of tests must include known positive and negative control samples to ensure that everything has worked properly. To examine a specimen from one patient requires that at least three tests are done. To avoid this waste they are usually performed in batches, with one set of controls tested per batch. This is why these tests are usually done at intervals, though the delay that results operates against one of the main advantages of these techniques.

29.4.2 Indirect tests

The natural consequence of an infection is an immunological response. This may involve the activation of B cells, T cells or both. A B-cell response results in the development of antibodies and these can be detected in a patient's serum (Box 29.4 and Figure 29.2). A number of tests have been developed to do this, and most of them also measure the quantity of any antibody present. 'Probe and label' tests can be used to detect antibodies as well as antigens. Certain basic principles govern the interpretation of the results of these tests.

Box 29.4 Immunoglobulin classes

Antibody, or immunoglobulin (Ig), comes in a variety of types or classes, with different functions.

IgA
Immunoglobulin of class A is designed to act after it has been secreted onto the surface of the body. It is specially active in the respiratory and gastrointestinal tracts, and it is also found in sweat, tears and breastmilk.

IgD
This immunoglobulin contributes to the activities of some lymphocytes.

IgE
IgE has a close relationship with mast cells and they can act together to produce allergic reactions.

IgG
This is the main, long-lasting antibody found in blood and tissues.

IgM
Class M immunoglobulins are large molecules of antibody that appear early in infections. They are very efficient activators of complement.

Antibodies specific for a particular microbe can be detected for the first time some days after the start of an infection. The speed and scale of the response varies from person to person, but the initial antibody is usually of the IgM class, followed by IgG (Figure 29.2). IgM antibody tends to disappear after a few months, but IgG persists, perhaps for life. Many tests do not distinguish between IgM and IgG antibody so do no more than confirm that there has been an infection with the related microbe at some time in the past. This may be in the last few days, or years ago. After an infection the amount of antibody declines with time but at speeds that vary with different infections and between individuals. The amount of antibody present is usually expressed as a **titre**. This is the reciprocal of the highest dilution of serum at which a test for antibody remains positive, so a maximum dilution of 1/128 becomes a titre of 128.

The best approach to the serological diagnosis of an infection is to collect two specimens of serum from the same patient about a week apart. This 'paired specimen' is examined for an increase in the amount of antibody (a rise in titre) to a particular pathogen between the two specimens. A four-fold rise (from 32 to 168, for example) or greater, is taken as significant. The logic

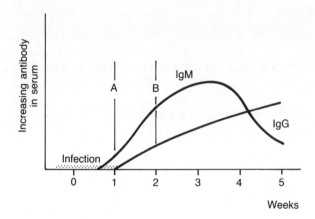

Figure 29.2 The development of antibodies in response to an infection. Antibody molecules of immunoglobulin class M (IgM) appear first but soon disappear, those of class G develop more slowly, and tend to persist (Box 29.4). The points marked A and B are separated by about a week. The increase in the amount of antibody between these points can be detected in the laboratory if a 'pair of sera' are tested, one of the pair collected at each point.

of this is that in most infections the titre of antibody rises from zero at its outset to reach a maximum in about two or more weeks (Figure 29.2). The results of this test may be of academic interest only, as many patients have recovered by the time the second sample has been collected.

To overcome this difficulty tests have been introduced that measure IgM antibody only, so a single specimen collected fairly early in the course of an infection can give a good indication of its cause. The minimum time before antibody can be detected is about seven days. As immunization begins during the incubation period antibody can sometimes be detected before the seventh day of an illness, as this is not the same as the seventh day of the infection.

The laboratory detection of cellular (T-cell) immunity is still a research procedure, but its presence or absence can be elicited clinically by the injection of a specific antigen into the skin, when one is available. The tuberculin test in tuberculosis is an example of such a test (Chapter 9).

FURTHER READING

See Appendix B.

The treatment of infections | 30

30.1 INTRODUCTION

In the nineteenth century deaths from infection were everyday facts of life (Box 30.1). Even in late Victorian Britain an average family with five or six children would be lucky to see four of them reach adulthood. At the time it was believed that much of this mortality was due to a lack of hygiene. Well over 100 years later infant and child mortality similar to that in early nineteenth-century Britain is still commonplace in some parts of the world. The most important cause is now identified as malnutrition. Lack of hygiene is an important secondary factor. Both are rooted in poverty and lack of education.

Box 30.1 Mortality in childhood

In the 15 years after they were married in 1860 a couple who lived in a rural part of the County of Kent, England, had 12 children. Six of them lived to reach adult life. Three (including twins) died in infancy, and the others at the ages of two, four and 15. A family history of this sort was quite usual at that time. Many of the deaths were the result of infections.

It did not become possible to mount a direct attack on most of the microbes that cause infections until the first effective antibacterial drugs appeared in the 1930s. Before this the only defences available were patients' own immune systems, so the outcome depended to a great extent on their general state of health. The contribution made by doctors and nurses was to provide support for patients in what was often a protracted struggle against the invaders. Skilled nursing was an important part of the process. When antimicrobial drugs appeared and infections were overcome in days rather than weeks the special skills developed over the years no longer seemed to matter. Many have been forgotten. This is a pity because infections have not gone away and victims still need skilled care.

In her *Notes on Nursing* (1859) Florence Nightingale not only underlined the central importance of cleanliness and fresh air, but also paid close attention to the rest of the environment in which patients are cared for. She stressed the need for bright and attractive surroundings, with a view, in which patients are kept occupied. She emphasized the avoidance of noise and sudden movements. Food was a major concern, both what it should be, and how to present it. She comments '. . . thousands of patients annually starve in the midst of plenty . . . ', and points out how this can be avoided. It is a salutary exercise to imagine what Miss Nightingale would think if she could visit hospitals today. It is fortunate that most patients spend so little time in them, and can escape to convalesce at home. Perhaps as they leave they should be given a booklet of 'hints', as Miss Nightingale called her *Notes*, to help them find the peace and quiet, and design the diet they need to speed their recovery (Nightingale, 1859; Grahame-Smith, 1993).

30.2 ANTIMICROBIAL THERAPY

Chemotherapy was the name applied to the treatment of infections with drugs produced by chemical synthesis. The first broad spectrum chemotherapeutic agents were the sulphonamides. Antibiotics are substances produced by one microbe that damage or destroy others. Penicillin was the first of these (Box 30.2).

Box 30.2 The first antimicrobials

The recorded history of antimicrobial therapy began in 1619. In that year an extract of cinchona bark (quinine) was used to treat the wife of the Spanish governor of Peru, who had malaria. She recovered. At about the same time emetine from ipecacuanha root was found to be useful in the treatment of some cases of dysentery, now known to be the amoebic variety. Much later arsenical drugs were used for the treatment of syphilis and in 1909 Paul Erlich introduced salvarsan for some protozoal infections. A major advance came in 1935 when the German Gerhard Domagk found a drug that acted against several bacteria. This was prontosil, the first sulphonamide. In 1928, in London, Alexander Fleming discovered pencillin. Nothing came of this until the early 1940s in Oxford when Howard Florey and Ernst Chain turned the discovery to practical use.

Since the 1940s several new chemotherapeutic drugs have been synthesized and many antibiotics have been discovered. At the same time the nomenclature has become confused. Some drugs that were originally antibiotics are now produced by chemical synthesis, so for example chloramphenicol has become a chemotherapeutic agent. Others, like ampicillin, are made by chemical modification of an antibiotic precursor, so are hybrids. More recently substances used for the treatment of malignant diseases have also been called chemotherapeutic agents. To avoid this confusion the term antimicrobial drug has been employed to describe all chemotherapeutic agents and antibiotics used in the treatment of infections.

The antimicrobials are a group of drugs that have a useful action against microbes that attack humans, or the animals or plants on which the human race depends. They include antibacterial, antiviral, antifungal and antiprotozoal agents, and for convenience, anthelmintics as well. These substances are tolerated by human tissues at levels that are toxic to their microbial targets. They are unique because they must act in the presence of two separate living organisms to damage or destroy one and, so far as possible, leave the other unharmed.

All antimicrobial drugs have at least two names. One of them is the trade name given to it by its manufacturer, under which they market it. This is why the same drug sold by different manufacturers or suppliers appears under different names. The other name is the generic or official one given to it by the regulatory authorities of each country in which the drug is sold. This ensures that within each country at least the same drug has the same official name, no matter who manufactures it. The trade and generic names of the same drug can look and sound quite different, and the names of different drugs sometimes sound similar, so accidents are possible. It is safer to use the generic names, and this has been done here.

30.2.1 The administration of antimicrobials

Antimicrobial drugs vary in the route and frequency of their administration, and in their toxicity. Some are made up to be applied locally in the form of solutions in various liquids or as creams, ointments, tablets, capsules, suppositories or pessaries. These topical formulations are designed for the treatment of superficial infections of the skin, eye, or the mucous membranes of the gastrointestinal or respiratory tracts, or of the vagina. They may be used in bacterial, fungal, viral or protozoal infections, or in worm infestations. The topical route of treatment may be supplemented or replaced by other routes when infections are severe, or the microbes responsible are not situated on or close to one of the surfaces of the body.

Some topical antimicrobial preparations have steroids added to them. These are designed to suppress the local inflammatory reaction at the same time as the antimicrobial attacks the organism responsible for it. This may be helpful when the inflammation is unnecessarily intense but some degree of inflammation is required for the resolution of an infection. In some situations the addition of steroids allows an infection to spread while it gives a false impression of benefit as the inflammation disappears. The drugs used in topical preparations vary in the extent to which they are absorbed through surfaces to which they are applied. Some of the more toxic should not be spread over large areas, or applied for long periods.

Several antimicrobial drugs are absorbed into the tissues in an active form when taken by mouth. The oral route is convenient as it is painless and patients can medicate themselves, though compliance may be poor. Those who have studied the matter suggest that, when unsupervised, only about half the patients who are prescribed oral medications take them as intended (Wright, 1993). The absorption of drugs from the gastrointestinal canal may be strongly influenced by the simultaneous presence of food. Some drugs are prescribed to be given before, and others after, meals. Athough this may cause inconvenience on drug rounds these instructions are important and should not be ignored. Drugs that alter the normal flora of the gut may cause diarrhoea, which may be severe. In some cases drugs are given by rectal suppository to be absorbed through the lining of the lower part of the gut, though this route can be unreliable.

Some drugs are given by injection either because they are not absorbed when taken by mouth, or because the condition of the patient rules out the use of the oral route. This may be in the early stages of a severe infection in which treatment is urgent or if the gastrointestinal canal is not working properly. When drugs are injected they irritate the tissues into which they are delivered. The less irritant may be given into the muscles, otherwise the intravenous route is used, and again this may be preferred in severe infections. Drugs may be injected intravenously in a single shot (a bolus) or by drip dissolved in larger volumes of liquid over variable periods of time. The choice depends on circumstances and the degree of local irritation caused by more concentrated solutions.

Drugs absorbed into the tissues are eliminated in various ways. Some are excreted into the urine through the kidneys, some through the liver into the bile, and some by both routes. They may be changed into inactive forms before this happens or they may be excreted with their antimicrobial activity intact. In some individuals drugs may precipitate allergic reactions.

The speed with which drugs are eliminated varies greatly. With most, it is rapid and doses must be repeated three or four times a day to maintain useful levels at the site of an infection. Routes of excretion may be blocked partly or completely by damage to, or disease of, the kidneys or the liver. If patients cannot eliminate antimicrobials normally the dose must be modified to avoid a toxic accumulation. Hepatic or renal damage may develop or its intensity change in the course of an infection. When this is possible patients are monitored from day to day to ensure that the dose of drug is neither too high nor too low. The level of antimicrobial present in the blood of a patient can be measured in the laboratory. Assays are done to control the dose of more toxic drugs, or to ensure that there is enough present when high levels are needed for successful treatment.

Antimicrobial drugs may be used in the treatment of an established infection, or as prophylaxis, to prevent one. Prophylactic antimicrobials are given to reduce the risk of an infection in medical or surgical procedures that might otherwise be followed by serious infections. Experiments have shown that to be effective the antimicrobial must be present in the tissues at the right level at the time the procedure is carried out, and for a few hours afterwards. Prophylaxis should begin just before an intervention starts, typically as part of premedication before surgery. If given earlier there is a chance that prophylaxis will fail if there is time for the patient's normal flora to change by encouraging microbes that are resistant to the drug used. Administration after the procedure for more than 24 or 48 hours does not contribute to the prevention of infection but it does add significantly to the population of resistant microbes in the hospital and the community.

30.3 ANTIBACTERIAL DRUGS

Many antibacterial drugs do little more than stop bacterial multiplication. These **bacteriostatic agents** depend heavily on the normal defence mechanisms of the body to kill and clear away the bacteria that, although inactivated by the drug, are still alive. If they are not dealt with in this way they return to full activity when the drug is withdrawn. **Bactericidal agents** do kill bacteria, but only when they are multiplying. Populations of bacteria always contain a proportion that are at rest. These resting bacteria are not affected by antimicrobial drugs, and the larger the population of bacteria exposed even to a bactericidal drug, the more survivors there will be. This is why all antimicrobial drugs need some help from the body's defences if the whole population of microbes present in an infection is to be destroyed.

30.3.1 Antibacterial activity

The antibacterial activity of an antimicrobial drug can be measured in the laboratory. The result is expressed as the **minimum inhibitory concentration** (MIC) of the drug against the microbe in question. One way to measure this is to make serial dilutions of the drug in tubes of a clear liquid culture medium. Each tube in a series of six or eight contains half the amount of drug present in the preceding one. All the tubes are then innoculated with a small, standard dose of the bacteria under test and they are incubated (Figure 30.1). Any bacteria that can still multiply despite the drug present do so, and tubes that contain them become cloudy. The end-point (MIC) is the minimum amount of antimicrobial that prevents (inhibits) the growth of the bacteria that were added, so the tube concerned remains clear. The bactericidal antimicrobials (penicillins, cephalosporins and aminoglycosides, for example) kill bacteria at the same time as their growth is inhibited. If the fluid

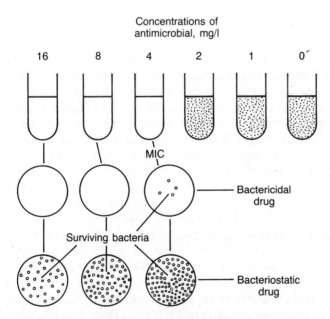

Figure 30.1 The measurement of the minimum inhibitory concentration (MIC) of an antimicrobial drug against a bacterium, and the distinction between a bactericidal and bacteriostatic agent. (See text; the tubes with growth are stippled. In the case illustrated the MIC is 4 mg/l.)

in the tubes that have remained clear are cultured, usually on a solid medium, only a few survivors are found. With the bacteriostatic antimicrobials (tetracycline, chloramphenicol, erythromycin) the clear tubes still contain most of the microbes that were added. Athough they have not multiplied, many are still alive. They begin to multiply again when the inhibitory antimicrobial is removed (Figure 30.1).

To succeed in its task in the body an antimicrobial must reach the tissues in the neighbourhood of the bacteria to be attacked at a concentration at least equal to the MIC of the drug for that organism. This explains why some bacteria reported as sensitive to an antimicrobial by a laboratory test do not respond to it in an infection. Most antimicrobials travel in the blood to reach the site of an infection. Treatment fails if the microbe is located in tissue with a restricted blood supply. To some extent this is the normal condition in bone, and it also applies to areas affected by arterial disease or where there has been vascular damage due to a prolonged inflammatory reaction. Some body sites are shielded from antimicrobials by natural anatomical or physiological barriers. Antimicrobials may not penetrate into the brain or the eye, or reach the insides of cells, so microbes in these places are protected. A laboratory report of sensitivity must be interpreted in the light of the clinical condition of the patient and the location and type of infection. Antimicrobial drugs may also fail in patients whose immune systems are deficient so they do not play their proper part in the destruction of the invader.

Some bacteria are naturally resistant to certain antimicrobials, or this resistance may be acquired. Acquired resistance implies that a bacterium that was originally sensitive has picked up a fragment of DNA that carries the information it needs to circumvent the action of the drug. The necessary fragment of DNA may be added ready-made by recombination from some other bacterium, or it may arise by alteration (mutation) of the bacterium's own DNA (Chapter 5). The mechanism of resistance may be an alteration in the permeability of the bacterial cell wall or cell membrane so that the anti-microbial is excluded from the cell. In some cases the chemical structure of the target for an antimicrobial is changed, so it is no longer effective. In other cases a bacterium produces an enzyme that destroys the antimicrobial before it can even reach its target. Two or more of these fragments of DNA may be acquired by a single bacterium, so it becomes resistant to several drugs at the same time. Any large population of microbes is likely to include a few that, by chance, have acquired one or more of these resistance mechanisms. If the population is then exposed to an appropriate drug the sensitive majority are wiped out, and the resistant minority multiply to fill the gap. In this way an apparently 'sensitive' population of bacteria becomes 'resistant'.

The range of bacterial genera and species susceptible to a particular drug is spoken of as its spectrum of activity. A **broad spectrum** drug is effective against a large number of different kinds of bacteria, a **narrow spectrum** one against only a few. If a broad spectrum drug is prescribed it will do more widespread damage to an individual's normal flora than one with a narrow spectrum. When the bacterial cause of an infection and its antimicrobial sensitivity are known, it is preferable to use a narrow spectrum drug to preserve as much as possible of the normal flora.

Four categories of target are available in bacteria for attack by antimicrobial drugs. These are indicated in Table 30.1, where the major groups of antimicrobials are classified according to their targets.

Table 30.1 A classification of some antibacterial drugs according to the points they attack in bacterial cells

Cell walls
The beta-lactams
　penicillins
　cephalosporins
　monobactams
　clavans
Vancomycin

Cell membranes
Colistin

Ribonucleic acid
The aminoglycosides
　streptomycin
　neomycin
　gentamicin
　tobramycin
　amikacin, etc.
Chloramphenicol
The tetracyclines
The macrolides
　erythromycin
　spiramycin, etc.
Fusidic acid
Rifampicin

Deoxyribonucleic acid
The sulphonamides
Trimethoprim
Metronidazole
The quinolones
　nalidixic acid
　norfloxacin
　ciprofloxacin, etc.

(a)　Antibacterial agents that act against cell walls

The peptidoglycan of bacterial cell walls is not found in human tissues. Any substance that can interfere with the production of peptidoglycan may have an antibacterial effect, without undue toxicity to a human host. Among the drugs that exploit this possibility are the bactericidal beta-lactam group of antimicrobials (penicillins, cephalosporins and monobactams) and vancomycin.

The formation of peptidoglycan begins inside the bacterial cell with the construction of the basic glycan building block. This is transported through the cell membrane by a carrier molecule, which then releases the glycan to join others to form long strands. The carrier molecule is recycled to repeat the process. Layer upon layer of glycan strands are built up and as this happens their peptide portions are cemented together to form a rigid structure just as bricks are joined together with mortar.

Vancomycin binds to and inactivates the carrier molecule. Penicillins and the other beta-lactam antimicrobials act to prevent the development of the vital peptide cement. The enzymes responsible for cementing the peptides together are situated on the cell membrane. Beta-lactam antimicrobials bind to these enzymes so they no longer work. There are a number of these so-called penicillin binding proteins (PBPs), each with a slightly different function. Different bacteria contain different permutations of PBPs and the various beta-lactam antimicrobials bind preferentially to different combinations of them. This explains some of the variations in the spectrums of activity of these drugs.

The drugs that attack the cell wall weaken it and eventually the bacterium ruptures like a punctured balloon because it can no longer withstand its own internal pressure. Acquired resistance to the beta-lactam antimicrobials may be due to changes in the molecular structure of the PBPs, to a change in the permeability of the cell wall that reduces the ability of the drug to reach them or to the production of enzymes (beta-lactamases or penicillinases) that destroy the drugs themselves. The beta-lactamases produced by different bacteria vary in their activity against various beta-lactam antimicrobials. This is another reason for the differences in the spectrums of activity of these drugs.

Benzyl penicillin (penicillin G) is the original narrow-spectrum penicillin. It has a high level of activity against several Gram-positive organisms, and against a few Gram-negative ones. It is very vulnerable to beta-lactamase, particularly that produced by *Staphylococcus aureus*. This has reduced its usefulness, but it is still the best drug for the treatment of organisms that have remained sensitive to it. It is unstable in acid, so cannot be taken by mouth. When given by injection it is rapidly excreted through the kidneys, so frequent doses are necessary. Excretion can be delayed and the number of injections reduced if less soluble compounds such as procaine or benzathine penicillin are used. If the drug probenecid is given at the same time the excretion of penicillin is delayed and very high tissue levels can be achieved. Skin rashes due to penicillin allergy are moderately common. Anaphylactic shock as a result of allergy is rare though it may be lethal. In general the drug is remarkably non-toxic, and high doses are well tolerated. Penicillin V is not destroyed by acids, so can be given by mouth.

Methicillin was the first penicillin to appear that was resistant to a beta-lactamase. It was used to treat infections with penicillinase-producing staphylococci though it has now been superseded by drugs like flucloxacillin that can be given by mouth as well as by injection. These drugs are less effective than benzyl penicillin against sensitive bacteria that do not produce beta-lactamases.

Broad-spectrum penicillins such as ampicillin are active against all the bacteria sensitive to benzyl penicillin, but two to four times as much of the drug is usually needed to produce the same effect. This loss of activity is compensated for by an extended spectrum that covers many Gram-negative species, though some of them also produce beta-lactamases to which the drug is vulnerable. Ampicillin is acid-stable, so can be given by mouth as well as by injection. It is not particularly well absorbed from the gastrointestinal tract, and residual ampicillin in the colon often disturbs the normal bacterial flora to cause diarrhoea.

Chemists have worked with the ampicillin molecule and changed it to produce a range of derivatives. These include amoxycillin (better absorbed from the gut, so less diarrhoea), azlocillin (powerful anti-pseudomonas activity, given by injection only), mezlocillin (generally better than ampicillin, by injection only), and piperacillin (combines the attributes of mezlocillin and azlocillin).

The core structure of all beta-lactam antimicrobials is now available in quantity, and a range of new drugs is likely to emerge as additions are made to it. The first 'monobactam' derived in this way is aztreonam which has good activity against most Gram-negative bacteria, and is resistant to nearly all the beta-lactamases they produce. It is inactive against Gram-positive organisms, and it is given by injection.

Another beta-lactam drug, imipenem, has a spectrum of activity that includes nearly all Gram-positive and Gram-negative bacteria, including the anaerobes. It is given by injection. It is destroyed by an enzyme found normally in the kidneys but this can be inactivated by the drug cilastin. The two are given together to prolong the life of imipeneum in the body, and to allow it to be used to treat infections in the urinary tract.

The cephalosporins are close relatives of the penicillins. As happened with the penicillins these have also been developed to give an increased spectrum of usefulness against Gram-negative species, though once more at the expense of a reduced activity against Gram-positive bacteria. Of the patients who are allergic to penicillin, 10% are also allergic to the cephalosporins. Cephalosporins are not very effective against intracellular bacteria because of poor penetration. Of the many that are available only a selection will be mentioned. Unless otherwise stated, they are given by injection.

Cefuroxime has useful activity against many Gram-positive and Gram-negative bacteria, but none against pseudomonas or the enterococci. It is resistant to many beta-lactamases. Cefuroxime axetil can be taken by mouth. Cefotaxime has a broader spectrum and is useful in bacterial meningitis. Ceftriaxone has an antibacterial spectrum similar to that of cefotaxime, but it is excreted slowly and doses of it may be spaced 12 or 24 hours apart. Ceftazidime is also similar to cefotaxime, but its spectrum extends to include *Pseudomonas aeruginosa*.

Clavulanic acid is the parent of a group of drugs called clavans. These inhibit the action of many beta-lactamases. They have little or no anti-microbial activity on their own but may be given at the same time as one of the broad spectrum penicillins. The clavans then act to neutralize any beta-lactamases present so the accompanying penicillin can reach its target before it is destroyed. Several mixtures of a clavan and a penicillin are on the market. Co-amoxyclav is a combination of clavulanic acid with amoxycillin.

(b)   Antibacterials active against cell membranes

Cell membranes are principally concerned with the control of what does or does not pass into or out of cells. Agents that act on the cytoplasmic membrane of bacteria are bactericidal though rather toxic to humans. Colistin is the only antibacterial agent in this class commonly used, usually only by the topical route.

(c)   Antibacterials active against ribonucleic acid (RNA)

Bacterial RNA exists in ribosomal, messenger and transfer forms. Ribosomes lie in the cytoplasm and they function to make protein. The information they need to do this originates in bacterial DNA and it is transmitted to the ribosomes in the form of a sequence of instructions carried on long strings of messenger RNA. Ribosomes slide along these strings and as they do so they join together the individual amino acids in the sequence prescribed by the messenger. The amino acids are brought to the ribosomes by transfer RNA. Rifampicin prevents the formation of bacterial messenger RNA, but does not affect its human counterpart. Bacterial but not human protein synthesis is prevented, and bacterial cells die.

The aminoglycosides (streptomycin, neomycin, gentamicin, tobramycin and amikacin) together with chloramphenicol, the tetracyclines, the macrolides (erythromycin, spiramycin), the lincomycins (lincomycin, clindamycin) and fusidic acid bind to various points on the ribosome itself. They all act to stop protein synthesis. Although the ribosomes of most bacteria are sensitive to these antimicrobials, permeability barriers sited at the cell wall exclude the drugs from some of them. For example, erythromycin does not penetrate into most Gram-negative bacteria, so cannot be used to treat the infections they cause.

The aminoglycosides are bactericidal, and the others, with exceptions for some bacterial species, are bacteriostatic. The aminoglycosides are rather toxic and quite minor overdoses may result in damage to the inner ear and the kidney.

(d)   Antibacterials active against deoxyribonucleic acid (DNA)

DNA is formed when large numbers of small, standard molecules are strung together to form long strands. Without these individual building blocks no DNA is made and cells deprived of it cease to multiply and begin to die. Bacteria (and other parasites) have to make their DNA from simple materials while humans can use the building blocks ready-made. The sulphonamides and trimethoprim interfere with the production of the basic blocks so are active against microbial parasites but are relatively non-toxic for humans. The sulphonamides are bacteriostatic; trimethoprim is bactericidal. They act at different points in the synthesis of the blocks so they may be more active when taken together. Cotrimoxazole is a mixture of these two drugs.

The quinolones (nalidixic acid, norfloxacin, ciprofloxacin, etc) inhibit an enzyme that acts to coil the very long molecules of DNA into neat packages that fit easily inside small organisms. Nalidixic acid is taken by mouth to treat urinary tract infections caused by Gram-negative bacteria. The newer fluorinated derivatives have a much enhanced activity, and a wide antibacterial spectrum. Some can be given by injection. Unfortunately bacteria rather easily develop resistance to them.

Metronidazole was originally develolped as an antiprotozoal agent. Later it was found to be very useful in the treatment of infections due to a wide range of anaerobic bacteria. Metronidazole is converted into an active form inside cells that are deprived of oxygen. It then breaks down DNA.

(e) Synergy between antibacterial agents

Two or more antimicrobial drugs act synergically if, when given together, they produce an effect that is greater than the sum of their independent actions. If the combined action is reduced below that of the more active component on its own they are antagonistic. If their combined effect does no more than to add together their independent actions the outcome is said to be additive. There are no hard and fast rules that predict into which of the three categories any particular combination will fall. To determine the outcome each mixture must be tried separately against the pathogen in question. This can be done in the laboratory but it is time-consuming. Experience has shown that the combination of an aminoglycoside with a penicillin is usually synergistic. If two bacteriostatic agents that act on the ribosome are used together they may get in each other's way, so be antagonistic.

30.4 ANTIVIRAL DRUGS

As with bacteria, viruses are only susceptible to antiviral drugs when they multiply. They use the metabolic systems of the cells they invade to achieve this so they are difficult to attack without unacceptable damage to the host. It follows that antiviral drugs tend to be toxic, and that they are virustatic rather than viricidal. To succeed antiviral drugs need to be backed up by effective host defence mechanisms.

Today's antiviral drugs attack virus multiplication at one of three points. The first is as they enter cells and are uncoated to release their nucleic acid. The second is when new viral nucleic acid is manufactured, and the third point is the production of the protein units used to form viral capsids (Chapter 16).

Table 30.2 Some antiviral drugs arranged by their points of action in the process of viral multiplication, and the principal viruses against which they are used

Antiviral agent	Used against
Viral entry into cells, and uncoating	
Amantadine	Influenza A
Synthesis of viral nucleic acid	
Acyclovir	Herpes simplex, varicella-zoster
Foscarnet sodium	Cytomegalovirus
Ganciclovir	Cytomegalovirus
Idoxuridine	Herpes simplex
Ribavirin	Respiratory syncytial virus, Influenza B, Lassa fever
Zidovudine (AZT)	Human immunodeficiency virus (HIV)
Synthesis of viral protein	
Interferon	Heptatitis

Some viruses have demonstrated an ability to become resistant to drugs to which they were once sensitive. Some of the antiviral drugs are summarized in Table 30.2. More are under development.

30.4.1 Viral entry and uncoating

This is the point at which amantadine acts against the influenza A virus. If given at the time of exposure to infection it may prevent disease altogether, or if given within the first 48 hours it reduces the severity of the illness. It may be used to protect especially vulnerable patients, for example the elderly or the debilitated. Its toxic side effects (nausea, dizziness, headache and insomnia) may be more severe than the influenza itself. It has no activity against other respiratory viruses.

30.4.2 Synthesis of viral nucleic acids

So far this has been the most fruitful point of attack for antiviral drugs. Idoxuridine is active against the DNA of the herpes simplex virus. It is too toxic for systemic use but may be applied locally to the dendritic ulcers that often develop when this virus attacks the eye, or to a herpetic whitlow.

Acyclovir has useful activity against the herpes simplex and varicella-zoster viruses. It is converted into its active form most freely inside cells that have already been infected, so it is less toxic than many other antiviral drugs. Acyclovir may be applied topically, taken by mouth or given by injection. It has changed herpes encephalitis into a treatable condition and if applied early it can lessen the severity of disease in cases of herpes simplex and herpes zoster. It has proved particularly useful in immunocompromised patients, including those with AIDS though the cytomegalo- and Epstein-Barr viruses are unaffected by it. A related drug, ganciclovir, is useful in the treatment of cytomegalovirus infections.

The drug ribavirin is more active against RNA viruses. It has been used in the treatment of respiratory syncytial virus infections in infants, in influenza B, and Lassa fever.

Zidovudine (azidothymidine, AZT) is used in the treatment of infections with the human immunodeficiency virus (HIV). It has been shown to slow down (though not halt) the progress of the disease in its late stage, when opportunistic infections begin to appear. It was hoped that it would also slow down progress in the early asymptomatic phase of the disease. This idea has been surrounded by some controversy and current research suggests that it does not happen. The drug is quite toxic. Some of its side effects are unpleasant (nausea, vomiting, muscle pains) while another, the suppression of the bone marrow, is dangerous. A number of drugs still under development may prove to be of value in the treatment of HIV infections. In due course these drugs may be more effective when sufficient are available to be used in combinations.

30.4.3 Synthesis of viral protein

The natural antiviral substance interferon (Chapter 2) is now available from commercial sources. So far it has not lived up to its early promise in the general treatment of viral infections. It has been used with some success to cut short the chronic stages of hepatitis B and C, also in some patients with AIDS.

30.5 ANTIFUNGAL DRUGS

It has been difficult to produce drugs active against fungi that are not unacceptably toxic to humans. Most of those in use act against fungal cell membranes, so share problems of toxicity with the antibacterial agent, colistin. The azole group (for example miconazole, ketoconazole and fluconazole) interfere with the manufacture of the membrane while amphotericin B and nystatin cause it to leak. Nystatin is only available for topical use, the others may be given systemically. Amphotericin B is rather toxic.

Griseofulvin is taken by mouth for severe dermatophyte infections. Flucytosine is sometimes used in serious generalized fungal infections. Both act on fungal DNA. No drugs have yet been producd that act against fungal RNA.

30.6 DRUGS ACTIVE AGAINST PROTOZOA AND HELMINTHS

It is not possible to produce a brief yet useful summary of the very wide range of drugs that have been developed for the treatment of infections and infestations with these parasites. Many of them are rather toxic. Some antibacterial agents (sulphonamides, trimethoprim, metronidazole, tetracycline) are also active against some protozoa. Readers in search of information about these drugs should seek it in more specialized works.

FURTHER READING

1. Nightingale, F. (1859) *Notes on nursing; what it is, and what it is not.* Reproduced in facsimile, 1974, Blackie, Glasgow.
2. Grahame-Smith, D.G. (1993) An encounter with Beethoven's cleaning lady. *Lancet*, **342**, 1315.
3. Wright, E.C. (1993) Non-compliance – or how may aunts has Matilda? *Lancet*, **342**, 909–13.
4. See Appendix B.

The Prevention of Infections

PART 9

Prevention in the community 31

31.1 INTRODUCTION

The health of a community is closely linked to the level of socio-economic development it has achieved. It has already been noted that infections are generally less severe among those with a high standard of living. In most cases they are also less common (Box 31.1). The factors that contribute

---

Box 31.1 High life, less infection?

The answer to this question is generally, yes, but some infectious diseases (as distinct from infections that may be subclinical) are more common in developed societies. In less favoured surroundings many very common infections are safely acquired in infancy, under the protection of maternal antibody. These infants do not suffer from disease (Chapter 2). With improved socio-economic conditions these infantile infections become less common and individuals meet them for the first time later in life, without the protection of maternal antibody. Disease rather than subclinical infection is then more often the result. Well established examples of these changes in the patterns of infection are poliomyelitis and hepatitis A (preventable by vaccines), plus herpes simplex and glandular fever. It is possible that bacterial meningitis is providing yet another.

---

to this are not easily disentangled, but some do stand out. Nutrition, housing and welfare are clearly important and another factor is the standard of medical care. To these are added certain more specific measures. These include vaccination, the enforcement of regulations that effect the hygiene of food and drink, the provision of a service for the surveillance of infections, and the handling of waste.

31.2 GENERAL FACTORS

By comparison members of an affluent society eat well, live comfortably, and are properly cared for in periods of special vulnerability, in infancy, pregnancy and old age. The relative importance of each of these components in the prevention of infection might be the subject of debate, but there is no doubt about the beneficial effect of an adequate, balanced diet.

Unfortunately what constitutes a diet that is adequate and balanced has become the subject of argument and some confusion. In recent years the spotlight has fallen, among other things, on the intake of roughage, fat, sugar and salt, and the special qualities of 'organic' foods. People naturally take an interest in what they eat. As the emphasis has changed from concern about where the next meal is to come from to what it is to contain, this interest may itself be unhealthy. Advice from 'experts' tends to be unreliable as dietary fads come and go and great conclusions are drawn from inadequate data. One thing that is certain is that the children who grew up in the 1940s in the UK when food was severely rationed were very healthy.

There is some truth in the idea that the health services in developed countries are really sickness services. More needs to be done to encourage a positive attitude to health, provided this does not itself lead to unhealthy introspection. Even so the existing level of medical and dental care contributes significantly to a reduction in the incidence of infection. Poor dental hygiene can effect nutrition and lead to tissue infections of the face and neck. This can be prevented by regular dental attention. Sufferers from diabetes and asthma form groups of significant size in the population. These and some other chronic diseases predispose to infections, more so when they are uncontrolled. Regular medical attention provides the control so prevents some infections.

31.3 VACCINATION

The immunological basis of immunization by vaccination has been described in Chapter 2. Here we are concerned with more practical matters. Vaccination has had an enormous impact on infection (Figure 31.1 and Box 31.2). Vaccination against smallpox has allowed this disease to be eradicated, and in many societies potentially lethal diseases such as diphtheria and poliomyelitis, once common, are now so unusual that a single case attracts considerable attention. As these and other serious infections have become rarities the interest of the public (or at least of the activists who purport to speak for them) has begun to shift from the effectiveness of vaccines to their safety.

When millions of apparently healthy people are vaccinated it is inevitable that a few will fall ill with some serious condition in the days that follow. If this leads to permanent disability or death the possibility that the vaccine was responsible is likely to be raised. Large sums of money may be paid in compensation if this can be proved in court so there has been a change in the atmosphere in which these questions are asked, and in the conclusions that are reached.

Some years ago there was wide publicity about a small number of cases in which whooping cough vaccine was alleged to have caused brain damage. Unbalanced and unenlightened discussion frightened the public to the extent that for a time a significant proportion of infants were not immunized against the disease. The result was an epidemic of whooping cough. Some infants who had not been immunized suffered permanent lung damage, and a

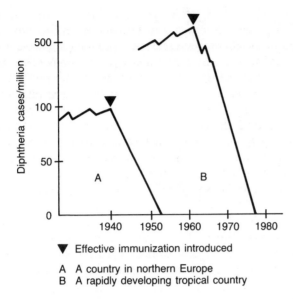

Figure 31.1 The effects of diphtheria vaccination on the incidence of the disease in two countries.

Box 31.2 Tuberculosis vaccine

Vaccination against TB has had a chequered history. When BCG vaccine became available in 1924 (Chapter 9) it was widely used in France. It was less well received in the rest of Europe and in N. America. There was concern about its safety, and the absence of proper controlled trials. In 1930 doubts about safety were heightened when there was a disaster in Lubeck in Germany. Over a quarter of some 250 infants given the vaccine died of tuberculosis. The cause seems to have been a batch of vaccine made with fully virulent tubercle bacilli instead of the BCG strain, but many people were badly frightened by the event.

Meanwhile in Scandinavia it was shown that nurses could be protected by the vaccine and between 1950 and 1963 a large trial was mounted in the UK. Among 50 000 school children the incidence of disease due to tuberculosis among those vaccinated was 4 per 10 000 while among the unvaccinated it was 19 per 10 000. This made it clear that although protection is not complete, BCG vaccination does give a useful level of protection.

---

number died. Some manufacturers were alarmed by the size of the compensation payments made and when they ceased to produce vaccines scarcities developed. Of course the safety of vaccines is an important issue, but no biological material can ever be entirely safe and a sense of proportion is needed when the matter is discussed.

When the efficacy and safety of vaccines have been dealt with satisfactorily, their stability and cost must be considered. All vaccines contain complex molecules that degenerate with time, and some contain whole living bacteria or active viruses. These molecules and microbes retain their full immunological activity for variable, usually short, periods of time. Most vaccines must be stored in a refrigerator if they are to remain active for long enough to be of use. In affluent societies this presents no real problem, though more attention needs to be paid to the storage temperature recommended (usually 4°C, without the risk of freezing). The temperature of any refrigerator in which vaccines are stored should be checked with an accurate maximum and minimum thermometer from time to time, or, when large quantities are involved, a recording thermometer and alarm should be fitted.

In less developed countries the lack of effective refrigeration has caused the failure of some vaccination programmes and has prevented the development of others. Many of these countries lie in the tropics. Vaccines deteriorate and become ineffective more quickly as the temperature rises. The answer is to establish a 'cold chain'. This consists of a series of depots equipped with refrigerators that continue to operate when the normal electricity supply fails, or that can work dependably in the complete absence of a public supply. These depots are spaced so that vaccines can travel between them in insulated containers by whatever transport is available and without an unacceptable

Table 31.1 Vaccines in common use, with an indication of the age at which they may be given. There are wide variations in practice between different countries both in the vaccines recommended and the ages at which they are used

Age	Vaccine	Notes
2–6 months	Diphtheria ⎫ Tetanus ⎬ Combined DTP Pertussis ⎭	Hepatitis B and yellow fever in some countries
	Poliovaccine (Sabin oral) (Salk injection vaccine preferred in some countries)	Salk vaccine for the immunocompromised
	H. influenzae type b (Hib)	
1–2 years	Measles ⎫ Mumps ⎬ Combined MMR Rubella ⎭	Measles at 6 months in Africa. Rubella in pre-pubertal girls if not immune
School entry	Boost diphtheria and polio	
School leaver	Boost tetanus and polio	
Elderly	Influenza, S. pneumoniae	Also for other high risk groups
Others	Rabies for those exposed or likely to be exposed Hepatitis B for high risk groups Varicella for the immuno-compromised	
Travellers	Yellow fever, typhoid, cholera, N. meningitidis Hepatitis A, Japanese encephalitis: according to proposed destinations	Boost yellow fever every 10 years and others as required

rise in temperature. Finally the vaccine must reach the health centres where it is to be used, perhaps carried on a bicycle or on foot, still within the cold chain. In the poorest countries where immunization is badly needed the combined cost of vaccines and their storage, transport and administration are powerful deterrents. The World Health Organisation and other bodies have channelled funds to try to meet this difficulty.

Recommendations for the use of vaccines vary from country to country, and programmes are kept under review to keep up with new developments. Table 31.1 lists the vaccines in common use, and the ages at which they may be given. All programmes represent generalizations and are devised with local conditions in mind. They may need to be adjusted in individual cases and in special circumstances. Vaccines may fail to produce immunity in the immunocompromised, in whom live vaccines can cause unacceptable reactions. Many vaccines must be given in multiple doses to achieve immunity in the first instance, and 'booster' doses are needed later, to sustain it. When a vaccine is first introduced into a community in which the associated disease is still common, the immunity of vaccinated individuals is boosted by occasional contacts with the real pathogen, so revaccinations may not be necessary. As the natural disease disappears these contacts need to be replaced artificially, by revaccination. Those who make and license vaccines take care to work out the intervals between doses and boosters to produce the optimum effect, so the timetables given in official recommendations or in the literature that accompanies vaccines should not be ignored.

31.4 HYGIENE OF FOOD AND DRINK

Across the world enteric fever (Chapter 10), cholera (Chapter 11), hepatitis A (Chapter 21) and food poisoning (Chapter 28) are all exceedingly common, and variably lethal. In more affluent areas enteric fever and cholera have virtually disappeared and hepatitis A has also become less common. These three infections are of human origin, and have a faecal-oral mode of transmission. The separation of human excreta from food and drink has been the key to their control. Where these diseases are still endemic or epidemic this separation has not been achieved. Enteric fever and cholera disappeared in countries where public health authorities have persuaded politicians to release the money necessary to build proper sewage systems and introduce piped supplies of clean water. Later the exercise of continued vigilance identified other loopholes through which excreta might still reach food and progressively these were blocked up. In northern Europe and North America most cases of enteric fever and cholera are now found in travellers who bring them back from other parts of the world.

It might seem strange that, in affluent countries, the same thing has not happened to food poisoning. The reason is that this infection has a different origin. So far it has proved impossible to achieve a separation between the most important of these, the excreta of infected animals, and food and drink. Although most of the sources of infection are known (or can be discovered) and despite the best efforts of armies of food inspectors, sanitary (health) inspectors, and veterinary and medical officials, supported to some extent by legislators, food poisoning has not gone away or even been reduced in incidence. There is a direct conflict between a desire for cheap, plentiful and convenient food and microbiological safety. Genuine and costly attempts have failed to eliminate salmonellas from poultry meat and eggs. The measures that would be needed to achieve this are, for the time being, commercially and politically unacceptable. The result is that large numbers of people still suffer from salmonellosis, and a few of them die of it.

31.5 THE SURVEILLANCE OF INFECTION

The standard definition of surveillance has overtones of spying and 'big brother' supervision, so the process has a bad name. The word does not adequately represent the activity to be discussed, which is benign in purpose and beneficial to the community. It might better be called surveyance which means the same thing without the overtones, though this word is found only in the largest dictionary. It is used here to make the point.

Nearly all benefits to the health of a community rest on the identification of a problem and the application of means to control it. The process is complete when the solution applied has been shown to work. The first and last of these processes involve surveys so are properly described as surveyance. As the towns and cities involved in the Industrial Revolution were developed, newly urbanized workers, including women and small children, were seriously exploited and they lived in squalid conditions. Records (surveys) of births and deaths showed the effect this had on mortality. This, together with the results of surveys of the health of the workforce, led to improvements that in the end benefited employees and employers alike. Other surveys confirmed the benefit, so the value of surveyance, and therefore of epidemiology, was established.

When the science of microbiology arrived it soon became clear that to produce a maximum effect on infection, epidemiology and microbiology could not be separated. Health services of all countries have recognized the importance of this joint approach, and those that can afford it have provided more or less elaborate and more or less effective systems to give it effect. The requirement is for a number of microbiology laboratories linked to an epidemiological service to form a national network that reaches into the grass-roots of health care in the community. Individual medical practitioners are usually required to feed epidemiologically important information into this network. The epidemiological and microbiological services collect, store and evaluate the data, provide resources for definitive diagnostic backup, and perform any special surveyance that is needed.

The importance of continued surveyance can be illustrated in several ways. The use of vaccines requires monitoring to ensure that they are, and remain, effective and safe, and that sufficient people are immunized to make an impact on the incidence of disease. Early warnings of danger or inadequacy can lead to remedial action before too much harm is done or money wasted. The fact that poultry meat and eggs are a (small) danger to the public health would be unrecognized without the epidemiological and microbiological skills of services whose task it is to conduct surveyance. That politicians do not yet know what to do with the information and are embarrassed by it does not detract from its value. The pandemic of AIDS was detected at an early stage by a service of this kind. Although a cure is still awaited the advanced information allowed provision to be made in time for the care of its victims. Legionellosis provides a similar though happier example. From surveyance and laboratory research came the information that now allows the disease to be diagnosed, treated and to some extent, controlled. These examples could be multiplied many times.

31.6 THE DISPOSAL OF WASTE

It is a common human perception that waste is not only offensive, but that it is also a hazard to health. This is strengthened when the waste arises in hospitals or other medical establishments. When the media report the finding of used syringes or human tissues in refuse tips or on beaches the matter becomes a major public issue. The attention paid to this special problem should not divert attention from the most important factor in the disposal of waste. This is the separation of sewage from drinking water. Separation must apply at the source of the water and be maintained throughout the system, including the places where pipes that contain each of them run together under the streets and into public buildings and homes.

The public perception that waste is potentially dangerous is correct in respect of excrement and some other items that may be disposed of in water closets. Faeces, and to a much lesser extent urine, may contain microbes or the eggs of larger parasites that can transmit the infections and infestations with which they are associated. Most other forms of waste are an aesthetic rather than an infectious problem. The appearance of such things as faeces, condoms and syringes, the latter usually flushed down closets by diabetics or drug addicts, has raised an outcry that demanded a response not least because used syringes may be dangerous if they fall into the wrong hands. The reason that remedial action has been slow is that it was difficult or impossible to prove that things other than syringes are really dangerous. This is why the easy and cheap solution, to discharge minimally or untreated sewage into the sea or into rivers, has in some places been retained for so long.

The type of waste with the worst reputation is that which emerges from hospitals and other medical establishments. In fact most of this is no more dangerous than that which emerges from the average home. About 10% is different, however, and this category may be described as 'clinical waste'. This includes used sharp objects (syringes and needles, scalpels, etc., mentioned further in Chapter 32), human tissues, used dressings and incontinence pads and some of the things disposed of by laboratories. Any of these have a potential for the transmission of infection if handled carelessly. Equally objectionable, though not for infectious reasons, are surplus or out-of-date drugs and other pharamceuticals, and confidential information about patients (Table 31.2).

Table 31.2 The categories of waste that arise in medical establishments, and their disposal. (With permission from Meers, P., Jacobsen, W. and McPherson, M. (1992) *Hospital Infection Control for Nurses*, Chapman & Hall, London, p. 125)

Categories	Disposal
Excrement and other liquid waste	Water carried sewage system, if available
Domestic rubbish: waste paper, plastic, kitchen waste, etc.	Same methods as are used in the local community for waste from households, with recycling if possible
Clinical waste: microbially hazardous waste,* human tissues	Incineration or such other methods as are approved by national authorities
Used sharp objects	Directly into special 'sharps' containers, then as above, incineration, etc.
Residues of drugs, confidential papers	As above, incineration, etc.
Radioactive, toxic or flammable materials	As directed by national authorities

*Used dressings, incontinence pads, disposable suction bottles, sputum pots, etc. Laboratory waste must be autoclaved before it leaves the department

Incineration of clinical waste is often the preferred option. Unfortunately this is costly and not everywhere environmentally acceptable. The usual alternative is controlled landfill but this may be even less acceptable to environmentalists and it is more subject to accident and error. Those who must choose between these options have a difficult decision to make. Much money would be saved if it were possible to separate medical waste into the 90% that can safely be disposed of in the same way as domestic waste, and the 10% that requires special provision. In practice this separation has proved difficult to achieve. Many hospitals have taken the safe, but expensive and environmentally unsound decision to treat all their waste as clinical waste.

FURTHER READING

See Appendix B.

Prevention in medical care | 32

32.1 INTRODUCTION

The last chapter dealt with the prevention of so-called community-acquired infections (CAIs), to be distinguished from infections that arise in hospitals (hospital-acquired infection, HAI). These are called nosocomial infections in some countries; the terms are synonymous. An HAI has been defined as an infection found in a patient in hospital that was not present and was not being incubated on admission, or one that was acquired in hospital but appeared after discharge from it. They include infections in members of hospital staffs acquired as a result of their employment.

It has been noted that real life tends to ignore neat human definitions and the distinction between CAI and HAI is no exception. Nowadays diagnostic and therapeutic interventions are often completed in the community because of the high cost of hospital care and admissions to hospitals are kept as short as possible. A single episode of illness may result in a number of brief admissions interspersed with periods of care elsewhere rather than in a single longer admission. (And hence the title of this chapter, 'Prevention in medical care' rather than 'in hospitals'.) These changes have made the simple definition of HAI more difficult to apply and less relevant. It might be logical to use the term iatrogenic to describe infections that are the result of therapy, no matter where this is carried out. This would introduce a new difficulty, however, as it would then be necessary to distinguish between infections that are related to treatment and those that are a part of a disease process. This problem does not arise when CAI and HAI are separated as the distinction depends on an objective geographical location rather than on a more subjective decision about the cause of an infection.

This is a problem for those whose task it is to record these infections. They need workable definitions if the data they collect are to be compared with those gathered at other periods and in other places. Fortunately our immediate concern is with the prevention of these infections, not with counting them. A great deal has been written about HAI and only a brief summary is provided here. For more detail the reader is referred to more specialized works (see Appendix B).

32.2 INFECTIONS DUE TO THERAPY

Florence Nightingale said hospitals should do the sick no harm. This absolute cannot be achieved. In the UK well over half of all deaths take place in hospitals. Most of these deaths are from the diseases for which individuals were admitted, but a few are the result of accidents that happen while they are in hospital. Only a minority of these accidents are fatal but all of them do some harm and are more or less avoidable. Accidents may be physical (a fall out of an unfamiliar high bed, too much or too little radiation); medical (diagnostic error, the wrong drug given, or in the wrong dose, or not given at all, the wrong operation performed, or the right one on the wrong side); psychological (stress due to admission to hospital or the treatment received), or the acquisition of an infection. Patients treated in the community may also suffer from several of these kinds of accidents. These are increasing in numbers as more of the patients who used to be cared for in hospitals are treated in the community.

Infections are inseparable from communal life, but they are more common and more severe in hospitals than elsewhere. Some people are admitted to hospital for the treatment of an established infection. The remainder are to some extent debilitated so even if not already infected they are more likely to acquire one. Because infections are more common many inpatients in acute hospitals are treated with antimicrobials. Surveys have shown that at any one time a third or more of patients are receiving these drugs, which may have been prescribed for the treatment of real (or imagined) infections, or as prophylaxis, to prevent them. The heavy use of antimicrobials selects bacteria that are resistant to those used regularly and these bacteria may also develop other undesirable characteristics that give them an advantage in hospitals. This is how 'hospital strains' of bacteria emerge and become involved in a vicious circle that turns hospitals into fertile seed-beds for infections. Debilitated patients are less able to defend themselves so their infections tend to be more severe, and those due to hospital strains may be more difficult to treat as well. The appearance of highly resistant bacteria is even more common in countries where the general public can purchase antimicrobial drugs over the counter, without a prescription.

The kinds of infections that make up CAI vary very much with the geographical location and degree of development of the communities concerned (Chapter 31). Differences in HAI are less marked because of international similarities in the way medicine is practiced. Data collected from many parts of the world show that urinary tract infections are usually the most common form of HAI, followed by infections of surgical wounds and of the respiratory tract.

HAI is not spread evenly through the hospital, either by specialty or by the type of infection (Figure 32.1). There is a close connection between the distribution of urinary catheters and the incidence of urinary tract infections (Table 32.1). Urinary catheters are used (or misused) more extensively in gynaecological practice, in certain surgical specialties and among the elderly, so patients admitted to these areas suffer more commonly from urinary tract infections. Of course surgical wound infections are seen most often in surgical wards, though nowadays because of early discharge and the growth of day surgery half or more of these infections in fact appear in the community. Respiratory infections are found most commonly in areas where people who require artificial ventilation are cared for, that is, in intensive care units.

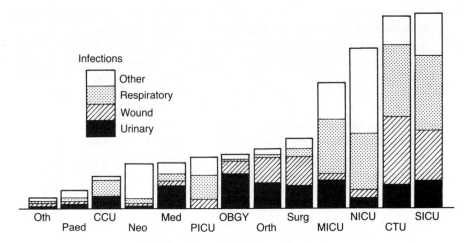

Figure 32.1 The amounts and types of hospital acquired infections found in different units and departments in the National University Hospital, Singapore, in the four years 1986–9. The heights of the columns are directly proportional to the incidence of infection in each unit.

Key: Oth, other departments; Paed, paediatric department; CCU, coronary care unit; Neo, neonatal unit; Med, general medicine; PICU, paediatric intensive care unit; OBGY, obstetrics and gynaecology; Orth, orthopaedic department; Surg, general surgery; MICU, medical intensive care unit; NICU, neonatal intensive care unit; CTU, cardiothoracic unit; SICU, surgical intensive care unit.

Reproduced with permission from Meers, P., Jacobsen, W. and McPherson, M. (1992) *Hospital Infection Control for Nurses*, Chapman & Hall, London p. 19, fig. 2.2.

Table 32.1 The association between urinary tract infections (UTIs) and the presence of urinary catheters. (Data from 18 163 patients surveyed in the UK National Prevalence Survey of Infection in Hospitals, 1980, see Meers, P.D. *et al.*, 1981)

Patients with	UTI	No UTI
Catheter in place	332*	1232
No catheter	477*	16 122

From the Table:

Patients infected and also catheterized, 21.2%
Patients infected but not catheterized, 2.9%

The probability that this observation might arise by chance is extremely small (the χ^2 statistical test gives a result of over 1000)

*Of these 809 infections 506 were acquired in hospital. UTIs made up 30.3% of all the infections acquired in hospital; it was the most common variety of HAI found

Other regular causes of infections are the intravascular catheters that are employed for diagnostic or therapeutic purposes. Such infections are naturally concentrated where intravascular devices are used most often and in particular where they are kept in place for longer periods.

As with the types of infection, so the microbes that cause them are unevenly distributed within hospitals. Infections of, or associated with, the surfaces of the body covered by skin are often due to staphylococci. Those of the urinary tract or the abdomen are more likely to be caused by Gram-negative bacteria from the gut. In very sick people the normal flora of the throat changes to include more Gram-negative species of bacteria so these are more often the cause of chest infections in patients in intensive care.

Careful studies in the USA have shown that about 6% of patients develop an infection as a result of admission to hospital. Among the patients who acquire these infections about 1% die as a direct result, and infection contributes to the death of another 2% of them. A patient who develops HAI and survives spends, on average, four extra days in hospital. In an ordinary general hospital this longer stay means that at any one time about 10% of patients are suffering from HAI (Box 32.1). HAI is very costly whether measured in terms of human suffering, or in money.

Box 32.1 Rates of infection

The fact that infection rates of 6% and 10% can exist simultaneously in the same hospital requires explanation. The 6% figure arises as a result of an **incidence survey**. The total number of infections (of HAI in this case) are counted ($\times 100$) and the product is divided by the number of patients exposed over the same period, say a year. The 10% figure comes from a **prevalence survey** in which every patient in the hospital is visited, preferably on the same day, and the number found with HAI ($\times 100$) is divided by the total visited. The prevalence figure is higher than the incidence one because patients with HAI spend about twice as long in hospital as those without. At any one time (= prevalence) more beds are occupied by patients with infections than is revealed by an incidence survey, which takes no direct account of the duration of a patient's admission.

Nurses and doctors working in hospitals often fail to notice that these infections are hospital-acquired. If asked about their daily experience they tend to say they do not see HAI (Box 32.2). The larger part of HAI (over 95%) appears continuously, at a fairly constant level; that is, it is endemic. This type of infection is less likely to be recognized for what it is. Outbreaks of HAI are much less common but much more obvious. Staff who under-react to the endemic form of the disease tend to over-react when an outbreak of HAI appears. The control measures that are then applied are often based on historical and erroneous ideas about infection that have not been proved to work. Sanctified by long use, many of these have become rituals.

Box 32.2 Perception of HAI

Only one-third of the nurses who attended a study day (and who had worked for an average of 10 years in hospitals) said they had seen a case of HAI as recently as within the last week. In fact it is probable that over 90% of them had done so. Half of them also thought that HAI is seen only as a part of an outbreak. The right message is not getting across.

A major study in the USA (called SENIC, the study of the efficacy of nosocomial infection control) has shown that a properly designed control programme can reduce the incidence of HAI by at least a third (Bennett & Brachman, 1992; Meers *et al.*, 1992). Nothing can happen, however, until hospital staffs recognize not only the existence, but also the scale of HAI. They need to react to the whole of it, not just to its epidemic part. For this reason programmes for the surveyance of HAI are necessary, not only to create awareness of the problem, but also to monitor the effectiveness of the methods used to control it.

The prevention of HAI produces several benefits. These include a reduction in pain, suffering and death among patients, and anxiety among their relatives. Individuals return to work more quickly. Patients spend a shorter time in hospital, use fewer drugs and have fewer investigations, so money is saved. More patients can be treated in the same number of beds, so efficiency is improved. Expenditure on infection control can be justified not only on humanitarian grounds, but also for sound economic reasons.

These worthwhile objectives can be achieved at least in part. This depends above all on the promotion of an awareness of the scale and importance of HAI. In the last 50 years hospital infection control has emerged as a minor medical speciality. Many hospitals have appointed specialists in this field, who have established infection control teams (ICTs), under the supervision of an infection control committee (ICC). The creation of awareness is the prime duty of ICTs and ICCs. An ICT is made up of one or more Infection Control Nurses (ICNs) with an Infection Control Doctor (ICD) who usually combines this duty with some other medical or surgical activity. Although these organizations were originally established for hospitals only, many have expanded their activities to take responsibility for the control of iatrogenic infections in the community.

ICTs and ICCs are available to give infection control advice whenever it is needed, and to react to problems with infections as they arise. ICTs operate systems of various sorts that are designed to detect changes in the patterns of infection or early indications of an outbreak. They should also monitor the impact of control programmes. Preventive measures depend to a considerable extent on the existence of adequate physical facilities such as those for the supply of safe food and of sterile equipment, and adequate provision for handwashing, for surgical operations, for any isolation of infected patients thought necessary, and for the diagnostic services. All of these provisions and services are monitored by ICTs. They should assist in the preparation of instructions for the performance of medical and nursing procedures that involve infection control measures and advise on the use of disinfectants. They can also help with the selection of medical and surgical equipment that might cause infections, and plan for its sterilization or disinfection. Another duty is in the field of education, within the area of their speciality.

These days in many countries every conceivable activity is regulated and documented. ICTs and ICCs have not escaped, and many have produced infection control manuals. These incorporate the policies and procedures agreed in hospitals and other medical facilities for which they are responsible. The circumstances in which these documents are written and used vary widely, and so do their contents. Opinions and practices differ so much that no useful overall review is possible outside the covers of more specialized works on HAI. Health professionals should familiarize themselves with the provisions for infection control that have been agreed in the organizations or institutions in which they work. In the event that such documents do not exist, or to expand on the information they contain, reference should be made to books on infection control (Appendix B). Only the subjects of disinfection and sterilization, and of the infections acquired by health care workers in the course of their duties, are sufficiently universal to be summarized here.

32.3   DISINFECTION AND STERILIZATION

For historical reasons the practices of disinfection and sterilization lie at the heart of infection control. When they were first introduced they were seen as the keys to the prevention of infection. As more has been learnt about the way infections are spread it can be seen that this was over-optimistic, but sterilization and to a lesser extent disinfection are still important topics. The methods commonly used in medical practice are illustrated in Figure 32.2.

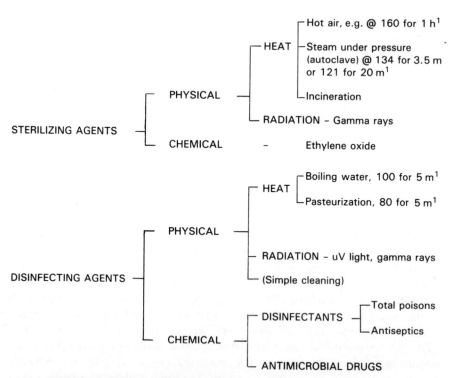

1. Other time and temperature combinations are used, for various purposes. Any combination chosen requires to have been fully validated, using heat-resistant spores or other microbes, as appropriate.

Figure 32.2 A summary of sterilizing and disinfecting methods commonly used in medical practice. Examples of temperatures are in °C, and times in minutes or hours. (Reproduced with permission from Meers, P., Jacobsen, W. and McPherson, M. (1992) *Hospital Infection Control for Nurses*, Chapman & Hall, London, p. 107)

The difference between sterilization and disinfection is that sterilization removes or kills all living matter, while disinfection implies the removal or destruction of sufficient microbes to make the object that is disinfected safe for the purpose intended. Of course if there are very few microbes present at the outset the process of disinfection may kill or remove them all so in fact sterility is produced, though this cannot be assumed. It is most important to ensure that whatever is to be disinfected (or sterilized) is thoroughly clean before the process is attempted. This makes the task easier and guards against failure.

32.3.1 Disinfectants

Bad smells are very often by-products of bacterial putrefaction. For this reason the human nose can detect the presence of certain kinds of bacteria, and this led to the use of antimicrobial substances before microbes were discovered. Certain chemicals were found to control smells, and because smells were associated with disease, they were also used to control infections. Chlorine and carbolic acid were employed in this way well over 100 years ago, and of course phenol was central to Joseph Lister's antiseptic surgery (Box 4.3). Although the microbes responsible for smelly putrefaction do not cause disease, most of those that do are also killed by these chemicals. Although the reason for their use was originally based on an error, the idea still works.

Numbers of other chemicals and mixtures of chemicals have joined these early disinfectants. It is important to note the general, direct relationship between the level of their toxicity to humans and their ability to kill or inactivate microbes. Very powerful disinfectants are extremely poisonous. Disinfectants that can be used on the skin or mucous embranes (also called antiseptics) are significantly weaker; indeed some more hardy bacteria are able to grow in dilute solutions of them. Infections have been caused in patients on whom supposedly 'sterile' disinfectants have been used. Powerful disinfectants must be used with care to avoid poisoning those who apply them, and they may do damage to some medical equipment or the materials of which furnishings and fabrics are made.

Microbes vary in their sensitivity to disinfectants. Only the most powerful have any useful activity against spores or *Mycobacterium tuberculosis*, and these require exposures more likely to be measured in hours than minutes. The hypochlorites and phenolics fall into this class, together with glutaraldehyde. Disinfectants (antiseptics) that can be used on the skin, such as the diguanides, triclosan and the quaternary ammonium compounds, are without useful effect against *M. tuberculosis*, spores or some viruses, and they act more quickly against Gram-positive bacteria than against some Gram-negative species. All these weak disinfectants are particularly easily inactivated by dirt, so anything to which they are applied must be thoroughly cleaned beforehand.

Detergents are weak antiseptics. A mixture of a detergent and an antiseptic can be used to clean and disinfect a dirty wound, in a single operation. A mixture of cetrimide and chlorhexidine is available for this purpose. Mixtures of detergents and antiseptics are also used when the hands are washed before surgery and in other clinical situations. Alcohol is the most powerful of the antiseptics. It evaporates when applied to the skin so only acts for a short time. Preparations of alcohol mixed with an antiseptic combine the rapid powerful action of one with the residual potential of the other. Alcohol is flammable, so must be used with care and kept away from flames or sparks, especially those from diathermy apparatus. When used it should be applied sparingly with a swab or sponge, or from a measured-dose dispenser. It should never be poured onto a patient from a bottle.

The volume of disinfectants used in hospitals has diminished greatly in recent years. They are now almost never employed for the treatment of floors, walls and other hard surfaces. Their use has been abandoned in many leading operating departments. No harm has been done, money is saved, and the environment is spared from toxic contamination. In the past when disinfection of the environment was still practiced it was also usual to sample it bacteriologically to monitor the disinfection process and to detect 'dangerous' contamination. This expensive and pointless activity has also been abandoned in most leading hospitals, again with no harm done. Antiseptics are a different matter, and they are still widely used for the disinfection of skin, mucous membranes and the tissues in wounds.

32.3.2 Sterilization

Many disposal items of medical equipment, for example syringes and needles, are sterilized by gamma radiation. The machinery required for this is large, complex and expensive, so the use of radiation is almost entirely confined to the commercial field. In ordinary health care facilities autoclaves or gas sterilizers are employed. Autoclaves use steam under pressure as the sterilizing agent. When well maintained and properly used, autoclaves produce a very high level of sterilization, with a wide margin of safety. Gas sterilizers use ethylene oxide, a toxic and potentially explosive gas. By comparison with autoclaves gas sterilizers are dangerous and have a narrow margin for error. Their only advantage is that they can sterilize objects that would be damaged or destroyed by the high temperature in an autoclave. They should only be used in well controlled surroundings, under expert supervision.

32.4 INFECTIONS IN HEALTH CARE WORKERS (HCWs)

By comparison with many industries, HCWs are employed in a very safe environment. Despite this many HCWs believe they are exposed to an increased risk of infection. Experience has shown that they are right, but that the risk is very small. No matter how small, however, employers and employees alike have a moral, and often legal, duty to minimize it. Good employers extend their commitment to patients to include the health of members of staff. In several countries legislation now makes this a statutory obligation. Employers must document the precautions introduced to prevent infections in their employees, and employees are required to apply them.

32.4.1   The hazard

Infection is a natural and so far unavoidable consequence of communal life. The risk of acquiring an infection is accentuated in hospitals and other health care establishments. To put this in perspective, most individuals suffer one or more minor infections each year. Sometimes these are more severe, particularly in childhood and old age. Among HCWs who have spent a lifetime in the industry very few have acquired a serious infection as a result of their employment. The more common minor infections are a hazard in any industry where employees have regular, close contacts with members of the public.

HCWs cannot fail to notice that a significant proportion of the patients they meet are under treatment for an infection. Many of these are complications of the illnesses for which they have come for treatment, or are unwanted consequences of that treatment. Very few of the microbes responsible are a threat to healthy HCWs, or healthy members of their families. An example is the methicillin-resistant *Staphylococcus aureus* (MRSA). Between 25 and 50% of all individuals carry 'ordinary' more sensitive *Staph. aureus* as a part of their normal flora, nearly always without knowing it. An HCW with a patient who carries or is infected with MRSA may pick it up and become a carrier of it. MRSA is more resistant, but so far no more pathogenic than other *Staph. aureus*, so the HCW is at no greater risk of serious disease than another healthy carrier of any staphylococcus. The only danger is that they may transmit it to the next patient they meet. As with any staphylococcus, this might cause an infection, but, with an MRSA, one that is more expensive to treat. In these circumstances it may be expedient to exclude the carrier from work until they are rid of their MRSA. This is a microbe that falls into a negligible category of risk for HCWs.

Not all microbes are associated with such a low level of risk. The infections they cause fall into two groups. In the first are those that are particularly dangerous to people (patients or staff) who are pregnant. Examples are chickenpox, shingles and rubella (Chapters 18, 20). HCWs may be protected from these as a result of natural infections they have suffered in the past, or by vaccination. If they are not protected in this way pregnant HCWs should be excluded from duties that are more likely to bring them into contact with these infections.

The second group of infections are those that can attack healthy people, with serious consequences. Infections with the hepatitis B virus (HBV) or the human immunodeficiency virus (HIV) fall into this class, and they are the subject of widespread concern (Chapter 21). Tuberculosis (Chapter 9) is another example.

HBV and HIV are spread by intimate activities that involve the transmission of body fluids, blood in particular, from one person to another. In the community they are spread almost exclusively by sexual intercourse or by intravenous drug abuse. When therapy is involved procedures that range from the use of hypodermic needles to major surgery offer artificial, highly effective routes for the transmission of HBV, and, to a lesser extent, of HIV. The reason for the difference is that the blood of a carrier of HBV contains many more infectious doses than is the case with a carrier of HIV (Chapter 21). Immunization provides a high level of protection against hepatitis B. Every HCW should be vaccinated against this disease and it is the responsibility of managements, through their occupational Health Departments, to see that this is done. So far there is no effective vaccine against HIV infection.

The problem with HBV and HIV infections is that many individuals are reluctant to admit, or are unaware, that they are carriers of one of these potentially dangerous viruses. The carrier status can only be revealed by laboratory tests. People under treatment for conditions unrelated to one of these infections are unlikely to be tested for them, and there has been resistance to the idea that all patients and staff should undergo these tests as a routine. HCWs must plan their activities in the knowledge that some of their patients and their colleagues may be carriers of these viruses. The frequency of this depends on the characteristics of the community from which patients and staff are drawn.

At present most HBV and HIV carriers remain undetected as they pass through health care facilities. The situation is complicated by the fact that even when the relevant laboratory tests have been ordered they cannot give instantaneous results, and an initial positive finding may be held back until it has been confirmed. This may take some days after patients have come under care and the appropriate specimens and requests have been submitted. The most dangerous period for HCWs is precisely this early period when most invasive diagnostic tests and therapeutic procedures are performed.

32.4.2 To contain the hazard

The importance of HBV vaccination has already been stressed. Apart from this HCWs can protect themselves if they adopt practices that minimize potentially dangerous exposures. A system called 'universal precautions' was developed in the USA to achieve this. Simple barriers, gloves for example, are used when exposure is expected to body fluids from any patient. Details of the technique are to be found in the infection control manuals of hospitals in which it has been adopted, or in more specialized works (Appendix B). The system has been criticized as expensive, and unnecessary in situations where the number of infected people is small. National health authorities or individual health facilities must decide, in the light of their circumstances, the precautions they wish to take.

There is less controversy about the other precautions that are necessary. These relate to the way in which clinical waste is handled (Chapter 31). The most important part of this is to reduce the number of accidents that result from the use and disposal of sharp diagnostic or therapeutic tools. Accidental cuts and punctures from sharp objects contaminated with the blood of patients are very common. These may involve both the primary users of these items or those who handle the containers of waste in which they are transferred to their final resting places. Although the attempt to prevent these injuries is not controversial, the methods that may be adopted to achieve it, are. The reasons for this are very practical, and they include genuine differences of opinion, and the shortage of money. Again local decisions must be taken in the light of local circumstances.

The other type of precaution that may be taken is to inform all HCWs who may be involved when a patient known to have a dangerous infection comes under care. They can then take any additional protective measures thought necessary. This approach may be modified by local rules concerning the confidentiality of this kind of information, but the procedure ought to be defined before it is necessary to use it. It must be stressed that the existence of such arrangements must not be allowed to give a false sense of security, or conceal the fact that most patients with these potentially dangerous infections will escape any special precautions because they are not identified in time, if at all.

Occasionally an HCW is discovered who is a carrier of HBV or HIV. If the duties of the individual include therapeutic activities that might lead to the infection of a patient, these must cease. Infected surgeons have been responsible for several transmissions of HBV to patients as a result of accidental cuts or punctures through gloves, during invasive procedures such as operations. Because of the larger volume of blood that must be transferred, the transmission of HIV in this way is less likely. There have been extremely few instances in which HCWs have infected their patients with this virus. Decisions in these matters may involve the future careers of expensively trained individuals so they are not taken lightly, but the peace of mind of a nervous public cannot be ignored. Many health authorities or other regulatory bodies have made recommendations that govern practice in this difficult area.

FURTHER READING

1. Bennett, J.V. and Brachman, P.S. (1992) *Hospital infections*, 3rd edn, Little, Brown, Boston.
2. Meers, P., Jacobsen, W. and McPherson, M. (1992) *Hospital Infection Control for Nurses*, Chapman & Hall, London.
3. Meers, P.D., Ayliffe, G.A.J., Emmerson, A.M. *et al.* (1981) Report on the national survey of infection in hospitals, 1980. *Journal of Hospital Infection*, **2** (supp), 23–8.

Appendix A
Microbial taxonomy

At the time the first microbes were described all other living things had been classified either as animals, or as plants. Microbes did not fit easily into these kingdoms, so a third, the **protists**, was proposed to accommodate them. A feature common to all three kingdoms is their basic unit, the cell. Cells may exist individually, or may be joined together to form more complex organisms. Viruses are insufficiently complex to be called 'cells' and may in any case be excluded from the classification because they are not really alive (Chapter 1).

The genetic blueprints of all organisms (excluding the viruses) are carried in molecules of **deoxyribonucleic acid (DNA)**. The DNA of all animal or plant cells is packaged within a membrane to form an obvious intracellular structure, the nucleus. This is called the **eukaryotic** condition, and organisms made up of nucleated cells are called **eukaryotes**. Bacterial DNA is not enclosed within a membrane, so although bacteria possess functional nuclei, they are not structural entities. This feature makes them into **prokaryotes**, and separates them from all living things including the other microbes, which are classified as eukaryotic protists. Viruses do not conform because although some have DNA as their genetic material others use a different molecule, **ribonucleic acid (RNA)**. All animals, plants and microbes (excluding the viruses) contain RNA as well as DNA. RNA functions to convert the information carried by DNA into the working parts of individual cells. Mature virus particles are quite different, as each of them contains either DNA or RNA, never both.

The scientific names of living organisms are given in Latin and are made up of two or more words (Chapter 1). The first word identifies a group of similar though not identical organisms that are allocated to the same **genus** and have the same first or **generic** name. The second or **specific** name is used to distinguish any separate types (**species**) that can be identified within a genus. Each of the types is given a different specific name. Sometimes a single species is subdivided into two or more subtypes or **varieties**. When this happens a third or fourth name is added. An example is the genus *Corynebacterium*. This is divided into several types or species to include those named *Corynebacterium xerosis*, *Corynebacterium hofmanii* and *Corynebacterium diphtheriae*. The species *Corynebacterium diphtheriae* includes a variety called *Corynebacterium diphtheriae* var *ulcerans*.

It is a convention that, when written down, these Latin words are printed in italic letters. In manuscript or typescript they are underlined to indicate that they ought to be in italic, if this were possible. The names of fungi, bacteria and protozoa may be given in their full, formal scientific format. For example, the bacterial cause of human tuberculosis is *Mycobacterium tuberculosis*, while the bovine form of the disease is caused by *Mycobacterium bovis*. Notice the capital initial letters for the generic and the lower cases for the specific names. Either may be referred to colloquially as a human or bovine tubercle bacillus, and both of them as tubercle bacilli. To refer to the genus with less formality but without using the colloquial expression one of them may be called 'a mycobacterium', or a number of them 'the mycobacteria' (no italics). The formal and informal generic names are the same, with allowances for singular and plural. The rules about Latin plurals are complicated!

Neisseria meningitidis, a cause of meningitis, is also a member of a genus that contains several species. When referred to formally they are collectively *Neisseria* spp., the 'spp.' being an abbreviation for the word species, here in the plural. When a single neisseria is mentioned without wishing to give its specific name the correct form is neisseria sp., the abbreviation in this case being in the singular. Informally as with 'mycobacteria' they may all be called neisseriae, though this awkward Latin plural may be anglicized to the more colloquial and easier 'neisserias'. In the same way *Salmonella* spp. may be called salmonellae or more easily the salmonellas. The last convention worth mentioning is designed to save space when repeatedly referring to the same or different species of a single genus. The first time the name is given in a paper, chapter or book it is written in full, for example *Neisseria meningitidis*. The next time a neisseria is mentioned the generic name is abbreviated. Practice varies a little in the degree of abbreviation allowed, but in this case on the second occasion it is often reduced to *N. meningitidis*, *N. gonorrhoeae*, or whatever the species may be.

There are some pitfalls to be noted. The word bacillus is used to describe any rod-shaped bacterium, but it is also the name of an important bacterial genus. The words bacterium or microbe may be used to indicate a single organism, or to describe any number, though collectively of the same type. Which meaning is intended is usually clear from the context.

Attempts have been made to give viruses latinized names, though this is not yet generally accepted. The virus of human herpes simplex may be called *Herpesvirus hominis*, but colloquial names such as, in this case, the herpes virus, or in other cases the influenza or measles viruses are more often employed. Journalists sometimes use the word virus when they discuss bacteria, so a 'cholera virus' or the 'virus' of Legionnaire's disease may appear. Although scientifically incorrect this error can be excused. The word virus described a poisonous or venomous material long before the first viruses were discovered.

As more has been learnt about the molecular structure of bacteria some older classifications based on simple tests have proved to be wrong. When new relationships are discovered the names of the microbes concerned may have to be changed. Another reason for a change is the application of strict rules to names that were originally formed incorrectly. It may seem that microbiologists alter the names of their menagerie to keep non-microbiologists guessing, but this is not so! Examples of newer names are *Clostridium perfrigens* for *C. welchii*, *Branhamella catarrhalis* for *Neisseria catarrhalis*, and *Enterococcus faecalis* for *Streptococcus faecalis*. The final authority in bacterial taxonomy is the *International Journal of Systematic Bacteriology*.

Appendix B
Further reading

A complete list of references to the primary sources for the statements made in this book would fill it several times over. Readers who need to trace original sources, or who wish to dig deeply into a particular topic, have several courses of action open to them. If they have access to a large medical library they may seek the information on the shelves, ask the librarian for help, or use whichever computer search facility is provided. Failing this the library in a centre for higher education or the reference section of a public library can assist. In these cases the response time may be slow and difficulties arise if multiple references are required.

Those who seek no more than confirmation, clarification or a general expansion of the information given here may refer to larger or more specialized textbooks, or to other sources. There are many of these, and readers may be limited by circumstances to whatever happens to be at hand. The books named below are trustworthy, and some provide detailed lists of references for more advanced students. Some of them are very expensive. The short list given does not imply that those excluded are of lesser merit, but simply that the present authors are familiar with the texts named, and have found them useful. Sources of more transient epidemiological information are also given.

Wide-ranging books

1. Jawetz, E., Melnick, J.L., Adelberg, E.A. *et al.* (1991) *Review of Medical Microbiology*, 19th edn, Appleton & Lange, Norwalk Conn. (concise, and regularly revised.)
2. Joklik, W.K., Willett, H.P., Amos, D.B. and Wilfert, C.M. (1988) *Zinsser Microbiology*, 19th edn, Appleton & Lange, Norwalk Conn. (A classic, sufficient for most purposes.)
3. Mandell, G.L., Douglas R.G. and Bennett, J.E. (1990) *Principles and Practice of Infectious Diseases*, 3rd edn, Churchill Livingstone, New York. (Compendious, many references.)
4. Benenson, A.S. (1990) *Control of Communicable Diseases in Man*, 15th edn, American Public Health Association, Washington. (An invaluable summary, regularly revised.)
5. Mims, C.A., Playfair, J.H.L., Roitt, I.M. *et al.* (1993) *Medical Microbiology*, Mosby Europe, London. (Concise, an interesting new approach with good illustrations.)

Books on special topics

6. Kucers, A. and Bennett, N. McK. (1987) *The Use of Antibiotics*, 4th edn, Heinemann Medical Books, Oxford. (Compendious, many references.)
7. Garrod, L.P., Lambert, H.P. and O'Grady, F. (1981) *Antibiotic and Chemotherapy*, 5th edn, Churchill Livingstone, Edinburgh.
8. *British National Formulary*, British Medical Association and the Pharmaceutical Society of Great Britain, London. (Revised every six months, includes useful notes on antimicrobial drugs and vaccines.)
9. Roitt, I.M. (1988) *Essential Immunology*, 6th edn, Blackwell Scientific, Oxford.
10. Bennett, J.V. and Brachman, P.S. (1992) *Hospital Infections*, 3rd edn, Little, Brown, Boston.
11. Ayliffe, G.A.J., Lowbury, E.J.L., Geddes, A.M. and Williams, J.D. (1992) *Control of Hospital Infection*, 3rd edn, Chapman & Hall, London.
12. Meers, P., Jacobsen, W. and McPherson, M. (1992) *Hospital Infection Control for Nurses*, Chapman & Hall, London.
13. Goldsmith, R. and Heyneman, D. (1989) *Tropical Medicine and Parasitology*, Appleton & Lange, Norwalk.
14. Burgess, N.H.R. and Cowan, G.O. (1992) *A Colour Atlas of Medical Entomology*, Chapman & Hall, London.

Other sources

15. Singleton, P. and Sainsbury, D. (1988) *Dictionary of Microbiology and Molecular Biology*, 2nd edn, John Wiley, Chichester. (Invaluable for the advanced student.)
16. *Communicable Disease Report**, Public Health Laboratory Service, London. Compiled weekly in the PHLS Communicable Disease Surveillance Centre, 61 Colindale Avenue, London NW9 5EQ, UK. (The CDR provides a continuous review of infectious disease in the UK, with frequent summaries and special reports.)
17. *Morbidity and Mortality Weekly Report**, Centers for Disease Control, Atlanta. Compiled weekly by the US Public Health Service, Centers for Disease Control, Atlanta, Georgia 30333, USA. (The MMWR does for the USA what the CDR does for the UK, but on a broader base of diseases.)

*Many other countries produce similar documents at weekly, monthly or other intervals. They are invaluable sources of up-to-date information about infectious diseases as they affect different parts of the world.

Index

Page numbers appearing in **bold** refer to figures and page numbers appearing in *italic* refer to tables.